T3-AKD-656

WITHDRAWN

On Sets Not Belonging
to Algebras of Subsets

Recent Titles in This Series

MEMOIRS
of the
American Mathematical Society

Number 480

On Sets Not Belonging
to Algebras of Subsets

L. Š. Grinblat

November 1992 • Volume 100 • Number 480 (third of 4 numbers) • ISSN 0065-9266

American Mathematical Society
Providence, Rhode Island

1991 *Mathematics Subject Classification.*
Primary 03E05, 04A20, 28A05, 54D35.

Library of Congress Cataloging-in-Publication Data

Grinblat, L. Š. (Leonid Š.), 1944–
   On sets not belonging to algebras of subsets/L. Š. Grinblat.
   p. cm. – (Memoirs of the American Mathematical Society, ISSN 0065-9266; no. 480)
   Includes bibliographical references.
   ISBN 0-8218-2541-0
   1. Combinatorial set theory. 2. Stone-Čech compactification. I. Title. II. Series.
QA3.A57   no. 480
[QA248]
510 s–dc20   92-28572
[511.3′22]   CIP

Memoirs of the American Mathematical Society

This journal is devoted entirely to research in pure and applied mathematics.

Subscription information. The 1992 subscription begins with Number 459 and consists of six mailings, each containing one or more numbers. Subscription prices for 1992 are $292 list, $234 institutional member. A late charge of 10% of the subscription price will be imposed on orders received from nonmembers after January 1 of the subscription year. Subscribers outside the United States and India must pay a postage surcharge of $25; subscribers in India must pay a postage surcharge of $43. Expedited delivery to destinations in North America $30; elsewhere $82. Each number may be ordered separately; *please specify number* when ordering an individual number. For prices and titles of recently released numbers, see the New Publications sections of the *Notices of the American Mathematical Society.*
   Back number information. For back issues see the *AMS Catalogue of Publications.*
   Subscriptions and orders should be addressed to the American Mathematical Society, P. O. Box 1571, Annex Station, Providence, RI 02901-1571. *All orders must be accompanied by payment.* Other correspondence should be addressed to Box 6248, Providence, RI 02940-6248.

Memoirs of the American Mathematical Society is published bimonthly (each volume consisting usually of more than one number) by the American Mathematical Society at 201 Charles Street, Providence, RI 02904-2213. Second-class postage paid at Providence, Rhode Island. Postmaster: Send address changes to Memoirs, American Mathematical Society, P. O. Box 6248, Providence, RI 02940-6248.

Table of Contents

ABSTRACT. This memoir is devoted to questions of the following type: Consider a set X with a family of algebras $\{\mathcal{A}_\lambda\}_{\lambda \in \Lambda}$ of subsets of X, $|\Lambda| \leq \aleph_0$. What conditions must be imposed on $\mathcal{A}_\lambda$ so that there exists a set not contained in any $\mathcal{A}_\lambda$?

Key words and phrases. Algebra of subsets, Čech compactification, ultrafilter, two-valued measure.

Wenn die Macht gnädig wird und herabkommt in's Sichtbare:

Schönheit heisse ich solches Herabkommen

Friedrich Nietzsche

1. INTRODUCTION

Without a doubt, after the construction of the classical (Lebesgue) theory of measure, some of the most interesting problems in set theory are those concerning nonmeasurable sets. Reflections on this range of problems led Banach to formulate the following general problem:

Problem. *Does there exist a real-valued function f, defined on all subsets of the interval $(0, 1)$, satisfying the following conditions:*

(1) *there exists a set M such that $f(M) > 0$;*

(2) *for every singleton $\{x\}$, $f(\{x\}) = 0$;*

(3) *if $\{M_n\}$ is a countable sequence of pairwise disjoint sets, then $f(\bigcup M_n) = \sum f(M_n)$.*

In 1930 Ulam proved the following theorem which partly solves this problem:

Theorem. *On the assumption that the continuum hypothesis is true $(2^{\aleph_0} = \aleph_1)$, there exists no function f satisfying the above conditions.*

Only forty years later it was shown that one can construct a model in which there exists a σ-additive extension of Lebesgue measure to all subsets of $(0, 1)$. This remarkable result is due to Solovay (see [So]). Ulam's result follows from his striking discovery, known as Ulam's matrix (see [U]):

Ulam's matrix. *If X is a set of cardinality $\aleph_1$, there exists a matrix of subsets of X:*

$$\begin{pmatrix} M_1^1 & M_2^1 & \ldots & M_\alpha^1 & \ldots \\ M_1^2 & M_2^2 & \ldots & M_\alpha^2 & \ldots \\ \cdots\cdots\cdots\cdots\cdots\cdots\cdots\cdots\cdots \\ M_1^n & M_2^n & \ldots & M_\alpha^n & \ldots \\ \cdots\cdots\cdots\cdots\cdots\cdots\cdots\cdots\cdots \end{pmatrix}$$

having $\aleph_0$ rows and $\aleph_1$ columns such that

(a) $M_\alpha^n \cap M_\beta^n = \emptyset$ *if* $\alpha \neq \beta$;

Received by editor March 26, 1990 and in revised form August 26, 1991

(b) $|X \backslash \bigcup_n M_\alpha^n| \leq \aleph_0$.

The existence of Ulam's matrix implies the following important corollary:

Corollary. *Let μ be a nontrivial σ-additive measure defined on a set of cardinality $\aleph_1$ ("nontrivial" means that $\mu(M) > 0$ for some M) such that $\mu(\{x\}) = 0$ for every singleton $\{x\}$. Then there exists $\aleph_1$ pairwise disjoint μ-nonmeasurable sets.*

This property of the cardinality $\aleph_1$ implies a well-known theorem of Alaoglu and Erdös, proved in [Er]:

Theorem. *Let $\mu_1, \ldots, \mu_k, \ldots$ be a countable sequence of two-valued measures[1] defined on a set of cardinality $\aleph_1$. Then there exists a set which is nonmeasurable relative to all these measures.*

We might mention that the special case of the Alaoglu-Erdös theorem for a finite sequence of measures was proved by Ulam. Moreover, the assumption that the measures are two-valued is not essential. The existence of Ulam's matrix implies that the Alaoglu-Erdös theorem remains in force provided that the measures μ_k are nontrivial, σ-additive, and that for every singleton $\{x\}$, $\mu_k(\{x\}) = 0$.

Alaoglu and Erdös arrived at their theorem as a result of their work on the following problem of Ulam:

Problem. *Find the minimal cardinal k such that, for any family of less than k two-valued measures defined on a set of cardinality $\aleph_1$, there exists a set which is nonmeasurable relative to all these measures.*

Ulam's problem was solved by Shelah (see[S]), who constructed a model in which $k = \aleph_1$. In Gödel's model L, however, $k = \aleph_2$.

We now proceed to formulate the main subject of this memoir. Our interest will be focused on sets which are not members of algebras of sets. Unless otherwise stated, all algebras here are defined on a certain abstract set X of arbitrary cardinality. By an *algebra* $\mathcal{A}$ we mean a collection of subsets of X possessing the following property:

---

[1] A measure μ defined on a set X is said to be two-valued if (1) μ is σ-additive, $\mu(X) = 1$, and $\mu(\{x\}) = 0$ for $x \in X$; (2) if M is a μ-measurable set, then either $\mu(M) = 0$ or $\mu(M) = 1$.

If M_1, $M_2 \in \mathcal{A}$, then $M_1 \cup M_2$, $M_1 \backslash M_2 \in \mathcal{A}$.

An obvious consequence of this property is that if M_1, $M_2 \in \mathcal{A}$, then $M_1 \cap M_2 \in \mathcal{A}$. We do not demand, unless otherwise stated, that singletons and X itself be elements of $\mathcal{A}$. If $X \in \mathcal{A}$, then $\mathcal{A}$ is commonly called a *field*[2] (see [Si]). By a set that is not a member of an algebra we mean, of course, a subset of X with that property.

Consider the following question: Can one construct, say, a finite sequence of algebras, none of which is all of $\mathcal{P}(X)$ [3], such that there is no set which is not a member of any of these algebras? If $\mathcal{A}_1$, A_2 are two algebras, none of which is all of $\mathcal{P}(X)$, then it is easy to prove the existence of a set which is a member neither of $\mathcal{A}_1$ nor of $\mathcal{A}_2$. However, if we consider three algebras, then the fact that none of them is all of $\mathcal{P}(X)$ does not yet imply the existence of a set which is not a member of any of them. Let X be a set of three distinct points x_1, x_2, x_3. Define three algebras $\mathcal{A}'_1$, $\mathcal{A}'_2$, $\mathcal{A}'_3$ by specifying all their members

$$\mathcal{A}'_1 \ni \emptyset, \{x_1\}, \{x_2, x_3\}, X;$$
$$\mathcal{A}'_2 \ni \emptyset, \{x_2\}, \{x_1, x_3\}, X;$$
$$\mathcal{A}'_3 \ni \emptyset, \{x_3\}, \{x_1, x_2\}, X.$$

Obviously, there is no set that is not a member of any of these three algebras. It is also obvious that for each of these algebras one can construct only two disjoint sets that do not belong to it.[4]

In light of this construction of algebras $\mathcal{A}'_1$, $\mathcal{A}'_2$, $\mathcal{A}'_3$, the following question – which is indeed the principal subject of this memoir – suggests itself quite naturally:

Question. Let $\mathcal{A}_1, \ldots, \mathcal{A}_k, \ldots$ be an at most countable sequence of algebras. What conditions must be imposed on these algebras so that there exists a set which is not a member of any of them?

We now formulate two results which are obvious corollaries of the two main theorems of this memoir:

[2] In the subsequent sections we will not use the term "field".
[3] By $\mathcal{P}(X)$ we denote, as usual, the set of all subsets of X.
[4] These arguments about two and three algebras are apparently not new.

Claim 1.1. *Let $\mathcal{A}_1, \ldots, \mathcal{A}_n$ be a finite sequence of algebras such that for every k, $1 \le k \le n$, there exist at most $\frac{4}{3}(k-1)$ pairwise disjoint sets which are not members of $\mathcal{A}_k$. Then there exists a set which is not a member of any $\mathcal{A}_k$, $1 \le k \le n$.*

Claim 1.2. *Let $\mathcal{A}_1, \ldots, \mathcal{A}_k, \ldots$ be a countable sequence of σ-algebras such that for every k there exist at most $\frac{4}{3}(k-1)$ pairwise disjoint sets which are not members of $\mathcal{A}_k$. Then there exists a set which is not a member of any $\mathcal{A}_k$.*

Let us return to the algebras $\mathcal{A}_1'$, $\mathcal{A}_2'$, $\mathcal{A}_3'$ constructed above. It follows from Claim 1.1 (and also from Claim 1.2) that if there existed not two but at least three pairwise disjoint sets not belonging to $\mathcal{A}_3'$, then there would exist a set not a member of $\mathcal{A}_1'$, $\mathcal{A}_2'$, $\mathcal{A}_3'$. This is no accident: We shall show that the bound $\frac{4}{3}(k-1)$ is in a certain sense the best possible.

Claim 1.2 is a generalization of the following theorem of Gitik and Shelah [GS]:

Theorem. *Let $\mu_1, \ldots, \mu_k, \ldots$ be a countable sequence of two-valued measures defined on a set of the power of the continuum. Then there exists a set which is μ_k-nonmeasurable for all k.*[5]

¿From the standpoint of the technique proposed in this memoir, the difficulty of Gitik and Shelah's theorem is not that one can construct a model in which there exists a two-valued measure μ, defined on a set of the power of the continuum, such that there do not exist $\aleph_1$ pairwise disjoint μ-nonmeasurable sets. The difficulty is as follows. One can construct a model (see [BD]) in which there exists a countable family of ultrafilters $\{a_\lambda\}$ over a set of the power of the continuum with the following property: define a set function μ by stipulating that $\mu(M) = 0$ if and only if $M \notin a_\lambda$ for all μ, and $\mu(M) = 1$ if and only if $M \in a_\lambda$ for all λ; then μ is a two-valued measure. Gitik and Shelah themselves used forcing to prove their theorem. A purely combinatorial proof was proposed by Fremlin and in [K]. We shall use the following equivalent version of the Gitik-Shelah theorem:

Theorem. *Let $\mu_1, \ldots, \mu_k, \ldots$ be a countable sequence of two-valued measures defined on a set of the power of the continuum. Then there exist pairwise disjoint sets $M_1, \ldots, M_k, \ldots$ such that M_k is μ_k-nonmeasurable.*

---

[5] This theorem is clearly a generalization of the Alaoglu-Erdös theorem – undoubtedly a profound and far from trivial generalization.

The proof of the equivalent of the Gitik-Shelah theorem and the last statement is very easy. Indeed, let $\mu_1, \ldots, \mu_k, \ldots$ be a countable sequence of two-valued measures defined on a set X of the cardinality of the continuum. By the Gitik-Shelah theorem, there exists a set M_1 which is μ_k-nonmeasurable for all k. On the set $X \backslash M_1$ we define two-valued measures $\mu_1^1, \ldots, \mu_k^1, \ldots$ as follows: $\mu_k^1(M) = 1$ if and only if there exists a set M' such that $\mu_k(M') = 1$ and $M = (X \backslash M_1) \cap M'$.[6] By the Gitik-Shelah theorem, there exists a set $M_2 \subset X \backslash M_1$ which is μ_k^1-nonmeasurable for all k. Now the measures $\mu_1, \ldots, \mu_k, \ldots$ induce two-valued measures $\mu_1^2, \ldots, \mu_k^2, \ldots$ on $X \backslash (M_1 \cup M_2)$. By the Gitik-Shelah theorem, there exists a set $M_3 \subset X \backslash (M_1 \cup M_2)$ which is μ_k^2-nonmeasurable for all k. Continuing in this way, we get a sequence of pairwise disjoint sets $M_1, \ldots, M_k, \ldots$ such that M_k is a μ_k-nonmeasurable set (in fact, for every n, M_n is a μ_k-nonmeasurable set for all k). On the other hand, let $M_1, \ldots, M_k, \ldots$ be pairwise disjoint sets, and let M_k be a μ_k-nonmeasurable set. For every n consider disjoint μ_n-nonmeasurable sets M_n', M_n'' such that $M_n = M_n' \cup M_n''$. Clearly, $\bigcup_n M_n'$ is a μ_k-nonmeasurable set for all k. $\square$

The author does not consider himself competent to describe the entire history of the main topic of this memoir. Suffice it to say that between the publication of Erdös' paper [Er] and Shelah's paper [S] there appeared perhaps two dozen publications in which the ideas of Ulam, Alaoglu and Erdös were developed. One of the more recent papers is that of Grzegorek [G] in which, as stated by the author in his abstract, he proves *a theorem generalizing results of Ulam, Alaoglu-Erdös, Jensen, Prikry and Taylor connected with Ulam's problem about sets of measures.* Grzegorek's theorem implies the following corollary:

Corollary. *Let $\mathfrak{F}$ be a family of σ-fields on the real line R such that for every $\mathcal{A} \in \mathfrak{F}$ all one-element subsets of R belong to $\mathcal{A}$ and $\mathcal{A} \neq \mathcal{P}(R)$. Then any of the conditions (i) $|\mathfrak{F}| < \omega$; (ii) $|\mathfrak{F}| < 2^\omega$ and $2^\omega = \omega_1$; (iii) $|\mathfrak{F}| < 2^{2^\omega}$ and Gödel's axiom of constructibility; implies $\cup \mathfrak{F} \neq \mathcal{P}(R)$.*

Claims 1.1 and 1.2, as stated above, are significant generalizations of Grzegorek's result in respect to conditions (i) and (ii).

---

[6] One usually says that the measure μ_k induces a measure μ_k^1 on the set $X \backslash M_1$. Obviously, μ_k will induce a measure on a set $L \subset X$ if and only if $\mu_k(L) \neq 0$.

Some remarks and notation. Unless otherwise stated, all algebras, measures and ultrafilters are defined on the same set X, which will be regarded as a topological space with the discrete topology. As usual, βX will denote the Čech compactification of X. The points of βX are ultrafilters over X. Ultrafilters will therefore be denoted by lower case letters. If $M \subset \beta X$ (in particular, if $M \subset X$), we let $\overline{M}$ denote the closure M in βX. The symbol $|M|$ will denote the cardinality of a set M.

The memoir consists essentially of two parts. The first part comprises Sections 4-9, the second part Sections 10-12. The main theorems will be stated in Section 2 and numbered, to set them apart from other theorems, by Roman numerals. Theorems I-IV and Theorem II* will be proved in the first part, Theorems V-XII in the second. It should already be clear from the formulations of these theorems in what sense the memoir splits into two parts. On the other hand, these two parts of our memoir combine quite naturally to form a unified whole – this will become particularly clear in the second part of Section 7. The basic idea of our method is outlined in Section 3. Section 13 may be regarded as an appendix to the memoir. It concerns what we shall call semi-lattices of subsets, and lattices of subsets, given, of course, on X. In Section 14 we list some unsolved problems which arose during our work on this memoir.

Acknowledgment. It is a pleasure to acknowledge the friendly support extended to me by M. Gitik throughout the lengthy research of which this memoir is the outcome, and his many useful comments. I am deeply grateful to D. Fremlin and the referee for their remarks.

2. Main Results

Theorem I. *(1) Consider a finite sequence of algebras $\mathcal{A}_1, \ldots, \mathcal{A}_n$ such that for every $k \neq 2$, $1 \leq k \leq n$, there exist more than $\frac{4}{3}(k-1)$ pairwise disjoint sets not in $\mathcal{A}_k$ (if $\mathcal{A}_2$ is taken into consideration, $\mathcal{A}_2 \neq \mathcal{P}(X)$). Then there exists $Q \notin \mathcal{A}_k$, $1 \leq k \leq n$.[7] (2) The bound $\frac{4}{3}(k-1)$ is best possible in the following sense: For every natural number $n > 1$ one*

---

[7] In order to give the reader an idea of the character of the material, we have stated two claims in the Introduction. The only difference between Claim 1.1 and the first part of Theorem I is that Claim 1.1 demands the existence of at least two pairwise disjoint sets not belonging to $\mathcal{A}_2$ (if $\mathcal{A}_2$ is taken into consideration). If $X \in \mathcal{A}_2$, then the statements of Claim 1.1 and the first part of Theorem I are identical.

can construct a sequence of algebras $\mathcal{A}_1, \ldots, \mathcal{A}_n, \mathcal{A}_{n+1}$ such that if $k \leq n$, there exist more than $\frac{4}{3}(k-1)$ pairwise disjoint sets not in $\mathcal{A}_k$, and such that there are only $[\frac{4n}{3}]$ [8] pairwise disjoint sets not in $\mathcal{A}_{n+1}$. Moreover, there does not exist a set which does not belong to all the algebras $\mathcal{A}_k$, $1 \leq k \leq n+1$.

Incidentally, our construction of the algebras $\mathcal{A}'_1$, $\mathcal{A}'_2$, $\mathcal{A}'_3$ in the Introduction actually yields a proof of part (2) of Theorem I in the case $n = 2$.

Before stating Theorem II, which is a generalization of Claim 1.2, we need the following

Definition 2.1. An algebra $\mathcal{A}$ is called an almost σ-algebra if for any countable sequence $M_1, \ldots, M_k, \ldots \subset X$, such that for any k, $\mathcal{P}(M_k) \subset \mathcal{A}$, it is also true that $\mathcal{A} \ni \bigcup_k M_k$.

Theorem II. *Consider a countable sequence of almost σ-algebras $\mathcal{A}_1, \ldots, \mathcal{A}_k, \ldots, \mathcal{A}_2 \neq \mathcal{P}(X)$, such that for every $k \neq 2$ there exist more than $\frac{4}{3}(k-1)$ pairwise disjoint sets not in $\mathcal{A}_k$. Then there exists $Q \notin \mathcal{A}_k$ for all k.*

As will be shown below (Example 6.1), Theorem II is no longer true if we do not demand that the algebras $\mathcal{A}_k$ be almost σ-algebras.

Theorem III. *(1) Consider an at most countable sequence of σ-algebras $\mathcal{A}_1, \ldots, \mathcal{A}_k, \ldots$ such that there exists a matrix of pairwise disjoint sets[9]*

$$\begin{pmatrix} U_1^1 \\ U_1^2 \\ U_1^3 & U_2^3 \\ \cdots\cdots\cdots \\ U_1^k & U_2^k \\ \cdots\cdots \end{pmatrix}$$

(each row from the third row on contains two sets) for which $U_i^k \notin \mathcal{A}_k$. Then there exists a set $U \notin \mathcal{A}_k$ for all k. (2) If not two but three rows of the matrix contain one set each, then the set U need not exist.

Part (2) of Theorem III was proved in the Introduction. The algebras $\mathcal{A}'_1$, $\mathcal{A}'_2$, $\mathcal{A}'_3$ constructed there are σ-algebras. There exist pairwise disjoint sets U_1, U_2, U_3 such that $U_i \notin \mathcal{A}'_i$, and there exists no set which is not a member of any of these algebras.

---

[8] $[\alpha]$ denotes the largest integer $\leq \alpha$.

[9] That is, a matrix of sets such that any two different sets in the matrix are disjoint.

8   L.Š. GRINBLAT

Theorem IV. *(1) Consider an at most countable sequence of algebras $\mathcal{A}_1, \ldots, \mathcal{A}_k, \ldots$. Suppose there exists a matrix*

$$\begin{pmatrix} U_1^1 \\ U_1^2 \\ \cdots \cdots \cdots \cdots \\ U_1^k \quad \cdots \quad U_{n_k}^k \\ \cdots \cdots \cdots \cdots \end{pmatrix}$$

of pairwise disjoint sets such that $U_i^k \notin \mathcal{A}_k$; $n_1 = n_2 = 1$; $n_k > 1$ for all $k > 2$; if $k \to \infty$, then $n_k \to \infty$. Then there exists a set $U \notin \mathcal{A}_k$ for all k. (2) If $k \to \infty$ but $\underline{\lim} n_k < \infty$, then the corresponding set U may not exist.

A particular case of Theorem IV is the following

Corollary 2.1. *Consider a finite sequence of algebras $\mathcal{A}_1, \ldots, \mathcal{A}_n$ such that there exists a matrix of pairwise disjoint sets*

$$\begin{pmatrix} U_1^1 \\ U_1^2 \\ U_1^3 \quad U_2^3 \\ \cdots \cdots \cdots \\ U_1^n \quad U_2^n \end{pmatrix}$$

(each row from the third row on contains two sets) for which $U_i^k \notin \mathcal{A}_k$. Then there exists a set $U \notin \mathcal{A}_k$ for all $k \leq n$.

Remark 2.1. Corollary 2.1 follows quite obviously from the arguments presented in the proof of Theorem III (see Remark 8.2).

Theorem V. *(1) Consider a finite sequence of algebras $\mathcal{A}_1, \ldots, \mathcal{A}_n$ such that for every $k \neq 2$, $1 \leq k \leq n$, there exist more than $\frac{5}{2}(k-1)$ pairwise disjoint sets not in $\mathcal{A}_k$ (if $\mathcal{A}_2$ is taken into consideration, $\mathcal{A}_2 \neq \mathcal{P}(X)$), and such that if $n > 1$, there exist three pairwise disjoint sets each of which is not a member of either $\mathcal{A}_1$, $\mathcal{A}_2$. Then there exist pairwise disjoint sets $V, U_1, \ldots, U_n$ such that if $U_k \subset Q$, $V \cap Q = \emptyset$, then $Q \notin \mathcal{A}_k$, $1 \leq k \leq n$. (2) The bound $\frac{5}{2}(k-1)$ is best possible in the following sense: For every natural number n one can construct a sequence of algebras $\mathcal{A}_1, \ldots, \mathcal{A}_n, \mathcal{A}_{n+1}^*$ such that for any k ($k \neq 2$, $1 \leq k \leq n$) there exist more than $\frac{5}{2}(k-1)$ pairwise disjoint sets not in $\mathcal{A}_k$; there exists $[\frac{5n}{2}]$ pairwise disjoint sets not in $\mathcal{A}_{n+1}^*$; and such that if $n > 1$, $\mathcal{A}_2 \neq \mathcal{P}(X)$, and there* exist three pairwise disjoint sets each of which is not a member of either $\mathcal{A}_1$, $\mathcal{A}_2$. Moreover, there do not exist corresponding sets $V, U_1, \ldots, U_n, U_{n+1}$.

Theorem VI. *Consider a countable sequence of σ-algebras $\mathcal{A}_1, \ldots, \mathcal{A}_k, \ldots, \mathcal{A}_2 \neq \mathcal{P}(X)$, such that for every $k \neq 2$ there exist more than $\frac{5}{2}(k-1)$ pairwise disjoint sets not in $\mathcal{A}_k$, and such that there exist three pairwise disjoint sets each of which is not a member of either $\mathcal{A}_1$, $\mathcal{A}_2$. Then there exist pairwise disjoint sets $V, U_1, \ldots, U_k, \ldots$ such that if $U_k \subset Q$, $V \cap Q = \emptyset$, then $Q \notin \mathcal{A}_k$.*

Theorem VII. *(1) Consider a finite sequence of algebras $\mathcal{A}_1, \ldots, \mathcal{A}_n$ such that for every k, $1 \le k \le n$, there exist more than $4(k-1)$ pairwise disjoint sets not members of $\mathcal{A}_k$. Then there exist pairwise disjoint sets $U_1, \ldots, U_n, V_1, \ldots, V_n$ such that if $U_k \subset Q$, $V_k \cap Q = \emptyset$, then $Q \notin \mathcal{A}_k$, $1 \le k \le n$. (2) The bound $4(k-1)$ is best possible in the following sense: For every natural number n one can construct a sequence of algebras $\mathcal{A}_1, \ldots, \mathcal{A}_n, \mathcal{A}_{n+1}$ such that for any k, $1 \le k \le n$, there exist more than $4(k-1)$ pairwise disjoint sets not in $\mathcal{A}_k$, and such that there exist $4n$ pairwise disjoint sets not in $\mathcal{A}_{n+1}$. Moreover, there do not exist corresponding sets $U_1, \ldots, U_n, U_{n+1}, V_1, \ldots, V_n, V_{n+1}$.*

Theorem VIII. *Consider a countable sequence of σ-algebras $\mathcal{A}_1, \ldots, \mathcal{A}_k, \ldots$ such that for every k there exist more than $4(k-1)$ pairwise disjoint sets not members of $\mathcal{A}_k$. Then there exist pairwise disjoint sets $U_1, \ldots, U_k, \ldots, V_1, \ldots, V_k, \ldots$ such that if $U_k \subset Q$, $V_k \cap Q = \emptyset$, then $Q \notin \mathcal{A}_k$.*

Theorem IX. *Consider a finite sequence of algebras $\mathcal{A}_1, \ldots, \mathcal{A}_n$. Suppose there exists a matrix of pairwise disjoint sets*

$$\begin{pmatrix} U_1^1 & \cdots & U_{m_1}^1 \\ \cdots\cdots\cdots\cdots\cdots \\ U_1^n & \cdots & U_{m_n}^n \end{pmatrix},$$

where $m_k > \frac{3}{2}(k-1)$, $U_i^k \notin \mathcal{A}_k$. Then there exists pairwise disjoint sets $V, U_1, \ldots, U_n$ such that if $U_k \subset Q$, $V \cap Q = \emptyset$, then $Q \notin \mathcal{A}_k$, $1 \le k \le n$.

Theorem X. *Consider a countable sequence of σ-algebras $\mathcal{A}_1, \ldots, \mathcal{A}_k, \ldots$. Suppose there* *exists a matrix of pairwise disjoint sets*

$$\begin{pmatrix} U_1^1 & \cdots & U_{m_1}^1 \\ \cdots\cdots\cdots\cdots\cdots \\ U_1^k & \cdots & U_{m_k}^k \\ \cdots\cdots\cdots\cdots\cdots \end{pmatrix},$$

where $m_k > \frac{3}{2}(k-1)$, $U_i^k \notin \mathcal{A}_k$. *Then there exist pairwise disjoint sets* $V, U_1, \ldots, U_k, \ldots$ *such that if* $U_k \subset Q$, $V \cap Q = \emptyset$, *then* $Q \notin \mathcal{A}_k$.

Theorem XI. *Consider a finite sequence of algebras* $\mathcal{A}_1, \ldots, \mathcal{A}_n$. *Suppose there exists a matrix of pairwise disjoint sets*

$$\begin{pmatrix} U_1^1 & \cdots & U_{m_1}^1 \\ \cdots\cdots\cdots\cdots\cdots \\ U_1^n & \cdots & U_{m_n}^n \end{pmatrix},$$

where $m_k > 3(k-1)$, $U_i^k \notin \mathcal{A}_k$. *Then there exist pairwise disjoint sets* $U_1, \ldots, U_n, V_1, \ldots, V_n$ *such that if* $U_k \subset Q$, $V_k \cap Q = \emptyset$, *then* $Q \notin \mathcal{A}_k$, $1 \le k \le n$.

Theorem XII. *Consider a countable sequence of* σ-*algebras* $\mathcal{A}_1, \ldots, \mathcal{A}_k, \ldots$. *Suppose there exists a matrix of pairwise disjoint sets*

$$\begin{pmatrix} U_1^1 & \cdots & U_{m_1}^1 \\ \cdots\cdots\cdots\cdots\cdots \\ U_1^k & \cdots & U_{m_k}^k \\ \cdots\cdots\cdots\cdots\cdots \end{pmatrix},$$

where $m_k > 3(k-1)$, $U_i^k \notin \mathcal{A}_k$. *Then there exist pairwise disjoint sets* $U_1, \ldots, U_k, \ldots,$ $V_1, \ldots, V_k, \ldots$ *such that if* $U_k \subset Q$, $V_k \cap Q = \emptyset$, *then* $Q \notin \mathcal{A}_k$.

We do not claim that the bound $\frac{3}{2}(k-1)$ in Theorem IX and the bound $3(k-1)$ in Theorem XI are best possible, though they correspond to such bounds in Theorems 10.5 and 10.6. Theorems IX and X are apparently true if $m_k > k-1$. Then, as is shown in Section 14 (Proposition 14.1), this bound is in a certain sense best possible. Theorems XI and XII are apparently true if $m_1 > 0$ and $m_k > 2k-1$ ($k > 1$). Then, as is shown in Section 14 (Proposition 14.2), this bound is in a certain sense best possible.

Remark 2.2. Obviously, the statement about the existence of sets $V, U_1, \ldots, U_n$ (in the first part of Theorem V, and in Theorem IX) is equivalent to the following statement: there exist pairwise disjoint sets $U_1, \ldots, U_n$ such that if $U_m \subset W \subset \bigcup_{k=1}^{n} U_k$, then $W \notin \mathcal{A}_m$. Thus, the statement about the existence of sets $V, U_1, \ldots, U_k, \ldots$ (in Theorems VI and X) is equivalent to the following statement: there exist pairwise disjoint sets $U_1, \ldots, U_k \ldots$ such that if $U_m \subset W \subset \bigcup_k U_k$, then $W \notin \mathcal{A}_m$.

We now turn to the last result, Theorem XII. The question that arises is, can one "refine" this theorem, that is to say, exploit the availability of the sets U_i^k to obtain information about the structure of a suitable sequence of sets $U_1, \ldots, U_k, \ldots$? The answer is in the affirmative. This will be done in Section 12, in Theorem 12.3. The same section will also contain improved versions of Theorems VI, VIII and X. Theorem 4.2 is a refinement of the first part of Theorem I. The refined versions of the first parts of Theorems III and IV follow immediately from their proofs and are formulated in Remarks 8.3 and 8.4. The first parts of Theorems V, VII, and Theorems IX, XI are refined in Section 10. We herewith present the most significant result of this memoir which is a refined version of Theorem II:

Theorem II*. *Consider a countable sequence of almost σ-algebras $\mathcal{A}_1, \ldots, \mathcal{A}_k, \ldots$. Suppose that for every k we are given a finite sequence of pairwise disjoint sets $Q_1^k, \ldots, Q_{m_k}^k \notin \mathcal{A}_k$, where $m_k > \frac{4}{3}(k-1)$ if $k \neq 2$. Then for every set of Q_i^k one can construct a set $\hat{Q}_i^k \subset Q_i^k$ such that $\hat{Q}_i^k \notin \mathcal{A}_k$; either $\hat{Q}_i^k = \hat{Q}_j^\ell$ or $\hat{Q}_i^k \cap \hat{Q}_j^\ell = \emptyset$; there exists a set Q which is the union of sets of type $\hat{Q}_i^p$ such that $Q \notin \mathcal{A}_k$ for all k; for every k there exists $\hat{Q}_{\psi_k}^{\xi_k} \subset Q$ such that $\hat{Q}_{\psi_k}^{\xi_k} \notin \mathcal{A}_k$.[10]*

All the refined versions mentioned above may be considered among the main results of this memoir.

3. Fundamental Idea

Consider an algebra $\mathcal{A}$ and $Q \notin \mathcal{A}$. There are two possible cases:

Case (1). If $Q \subset D$, then $D \notin \mathcal{A}$.

Case (2). There exists $D \supset Q$ such that $D \in \mathcal{A}$.

---

[10] The last assertion, concerning the existence of the sets $\hat{Q}_{\psi_k}^{\xi_k}$, is not trivial in order that $\mathcal{A}_k$ be an almost σ-algebra rather than a σ-algebra.

Consider the first case. Let $Q_1, \ldots, Q_m$ be a finite sequence of sets and $Q = \bigcup_{i=1}^m Q_i$. It is clear that there exists Q_i such that if $Q_i \subset M$, then $M \notin \mathcal{A}$.

As for the second case, let $Q_1, \ldots, Q_m$ and $Q'_1, \ldots, Q'_n$ be two finite sequences of sets, $Q = \bigcup_{i=1}^m Q_i$, $D \backslash Q = \bigcup_{i=1}^n Q'_i$. It is a simple matter to prove that there exist Q_i, Q'_j such that if $Q_i \subset M$, $Q'_j \subset M'$, $M \cap Q'_j = \emptyset$, $M' \cap Q_i = \emptyset$, then $M, M' \notin \mathcal{A}$. Obviously, if we replace the condition $D \backslash Q = \bigcup_{i=1}^n Q'_i$ by the requirement that $X \backslash Q = \bigcup_{i=1}^n Q'_i$, we can find appropriate sets Q_i, Q'_j.

It is a remarkable fact that these arguments directly imply essentially new results. To illustrate, we are going to prove a weakened version of the first part of Theorem I:

Proposition 3.1. *Consider a finite sequence of algebras* $\mathcal{A}_1, \ldots, \mathcal{A}_n$ *such that for every* k, $1 \leq k \leq n$, *there exist more than* $2(k-1)$ *pairwise disjoint sets not in* $\mathcal{A}_k$. *Then there exists* $Q \notin \mathcal{A}_k$, $1 \leq k \leq n$.

Proof. The proof proceeds by induction. It will suffice to prove the following:

Consider algebras $\mathcal{A}_1, \ldots, \mathcal{A}_k, \mathcal{A}_{k+1}$. *Suppose there exist a set* $R \notin \mathcal{A}_i$, $1 \leq i \leq k$, *and pairwise disjoint sets* $K_1, \ldots, K_\ell$, $\ell > 2k$, $K_i \notin \mathcal{A}_{k+1}$, $1 \leq i \leq \ell$. *Then there exists* $L \notin \mathcal{A}_i$, $1 \leq i \leq k+1$.

Indeed, consider a finite partition of X, say $Q_1, \ldots, Q_m$ ($X = \bigcup_{i=1}^m Q_i$, and $Q_i \cap Q_j = \emptyset$ if $i \neq j$), such that $R, K_1, \ldots, K_\ell$ are unions of sets Q_i. There exist

$$Q_{i_1}, \ldots, Q_{i_k} \subset R$$

(not necessarily different) such that for any $p \leq k$ at least one of the following two conditions holds:

(1) if $Q_{i_p} \subset D$, then $D \notin \mathcal{A}_p$;

(2) there exists $Q_{j_p} \subset X \backslash R$ such that if $Q_{i_p} \subset M$, $Q_{j_p} \subset M'$, $M \cap Q_{j_p} = \emptyset$, $M' \cap Q_{i_p} = \emptyset$, then $M, M' \notin \mathcal{A}_p$.

Since there are at most $2k$ sets Q_{i_p}, Q_{j_p}, while $\ell > 2k$, there exists K_r such that $K_r \cap Q_{i_p} = \emptyset$ for all $p \leq k$, and $K_r \cap Q_{j_p} = \emptyset$ for all sets of the form Q_{j_p}. If $\mathcal{A}_{k+1} \not\ni \bigcup_{p=1}^k Q_{i_p}$, then the required set is $L = \bigcup_{p=1}^k Q_{i_p}$. Otherwise, $L = K_r \cup \bigcup_{p=1}^k Q_{i_p}$. $\square$

When dealing with finite sequences of algebras (e.g., in the proof of the first part of Theorem I), quite elementary arguments, involving finite partitions of sets, will be sufficient. To deal with countable sequences of algebras, however, we shall have to speak in the language of ultrafilters.

We return now to the two cases distinguished at the beginning of this section. In Case (1) there exists an ultrafilter a such that $Q \in a$ and if $M \in a$, then $M \notin \mathcal{A}$. In Case (2) there exist ultrafilters a, b such that $Q \in a$, $D \backslash Q \in b$, and if $M \in a$, $M' \in b$, $M \notin b$, $M' \notin a$, then $M, M' \notin \mathcal{A}$.

Definition 3.1. Consider an algebra $\mathcal{A}$, an ultrafilter a is said to be $\mathcal{A}$-special if $M \in a$ implies $M \notin \mathcal{A}$. Different ultrafilters a, b are said to be $\mathcal{A}$-similar if there exists $D \in \mathcal{A}, a, b$, and whenever $M \in a$, $M' \in b$, $M \notin b$, $M' \notin a$, it follows that $M, M' \notin \mathcal{A}$.

Thus a is $\mathcal{A}$-special iff $a \cap \mathcal{A} = \emptyset$; and a, b are $\mathcal{A}$-similar iff they are distinct and $a \cap b \supset (a \cup b) \cap \mathcal{A} \neq \emptyset$.

If a, b are $\mathcal{A}$-similar ultrafilters, we say that a has an $\mathcal{A}$-similar ultrafilter b.

A necessary and sufficient condition for the existence of an $\mathcal{A}$-special ultrafilter is that $X \notin \mathcal{A}$.

$\mathcal{A}$-similarity of ultrafilters is a transitive relation: if a, b and b, c are pairs of $\mathcal{A}$-similar ultrafilters and $a \neq c$, then a, c are also $\mathcal{A}$-similar.

We now proceed to an exposition of the main idea of this memoir.

Proposition 3.2. *Let $\{\mathcal{A}_\lambda\}$ be a family of algebras.*[11] *A necessary and sufficient condition for the existence of a set that is not a member of any algebra in the family is that there exist sets $S, T \subset \beta X$ such that $\overline{S} \cap \overline{T} = \emptyset$ and for every λ at least one of the following two conditions holds:*

(1) there exists an $\mathcal{A}_\lambda$-special ultrafilter $z_\lambda \in S$;

(2) there exist $\mathcal{A}_\lambda$-similar ultrafilters s_λ, t_λ such that $s_\lambda \in S$, $t_\lambda \in T$.

Proof. If there exists a set which does not belong to all $\mathcal{A}_\lambda$ (denote it by Q), then there exist corresponding sets S, T and $S \subset \overline{Q}$, $T \subset \overline{X \backslash Q}$. Therefore $\overline{S} \cap \overline{T} = \emptyset$. On the other

---

[11] We stated in the Introduction that we are considering at most countable families of algebras. Here, however, we are speaking of a family of algebras of arbitrary cardinality.

hand, if there exist corresponding sets S, T and $\overline{S} \cap \overline{T} = \emptyset$, then we take $Q \subset X$ such that $S \subset \overline{Q}, T \subset \overline{X \backslash Q}$. It is clear that $Q \notin \mathcal{A}_\lambda$ for all λ. $\square$

Thus, the question of finding a set not contained in any algebra $\mathcal{A}_\lambda$ is a question regarding the "geometry" of βX or, one might say, the "combinatorics" of βX.

The following definition will be important in the sequel:

Definition 3.2. Consider an algebra $\mathcal{A}$, the set

$$\{a \in \beta X \mid \text{ either } (1) \ a \text{ is an } \mathcal{A}\text{-special ultrafilter};$$
$$\text{or } (2) \ a \text{ has an } \mathcal{A}\text{-similar ultrafilter}\}$$

is called the kernel of $\mathcal{A}$ and denoted by $\ker \mathcal{A}$.[12]

Let $1 \le k \le \aleph_0$. A necessary and sufficient condition for the existence of k pairwise disjoint sets not members of $\mathcal{A}$ is that $|\ker \mathcal{A}| \ge k$.

The following definition will prove useful in our further arguments:

Definition 3.3. A set $M \subset \ker \mathcal{A}$ is said to be $\mathcal{A}$-similar if $|M| > 1$, any two distinct ultrafilters in M are $\mathcal{A}$-similar, and there exist no $\mathcal{A}$-similar ultrafilters a, b such that $a \in M, b \in \beta X \backslash M$.

In the sequel we shall use the following two obvious assertions without special mention:

Assertion (1). *The kernel uniquely determines the algebra. This means the following: Let $\mathcal{A}$ and $\mathcal{B}$ be algebras, and suppose that: a is an $\mathcal{A}$-special ultrafilter iff a is a $\mathcal{B}$-special ultrafilter; a, b are $\mathcal{A}$-similar ultrafilters iff a, b are $\mathcal{B}$-similar ultrafilters. Then $M \in \mathcal{A}$ iff $M \in \mathcal{B}$.*

Assertion (2). *Let $\beta X \supset K \ne \emptyset$, $|K| < \aleph_0$; $M_1, \ldots, M_n$ are pairwise disjoint subsets of K, $K = \bigcup_{i=1}^{n} M_i$, and $|M_i| > 1$ if $i > 1$. It is possible that $M_1 = \emptyset$, and it is possible that $n = 1$. Each ultrafilter from M_1 is called $\mathcal{A}$-special, and each from sets M_i, $i > 1$, is called an $\mathcal{A}$-similar set. Then, in a natural way, one defines an algebra $\mathcal{A}$ with a corresponding kernel.*

---

[12] As is shown in Section 13, $s \in \ker \mathcal{A}$ iff there exists a set $Q \in s$ such that $(s \cap Q) \cap \mathcal{A} = \emptyset$.

4. FINITE SEQUENCES OF ALGEBRAS (1)

As mentioned in the previous section, to deal with finite sequences of algebras it is actually sufficient to consider only finite partitions of sets. Nevertheless, we find it more convenient to adhere to the language of ultrafilters even in this case.

Consider a finite sequence of algebras $\mathcal{A}_1, \ldots, \mathcal{A}_n$. According to Proposition 3.2, a necessary and sufficient condition for the existence of a set not in any of these algebras is that there exist two sets – which we shall denote throughout by S_n, T_n – whose elements are ultrafilters, with the following properties. Let S_n denote the set

$$\{s_1, \ldots, s_i, \ldots, s_n\},$$

where possibly $s_i = s_j$, $i \neq j$. Either the ultrafilter s_i is $\mathcal{A}_i$-special or else we consider a pair s_i, t_i of $\mathcal{A}_i$-similar ultrafilters. Define T_n as the set of all t_i. There may exist i, $1 \leq i \leq n$, for which T_n contains no t_i; it is also possible that $t_i = t_j$, $i \neq j$. We do stipulate that $S_n \cap T_n = \emptyset$. By Proposition 3.2, we must demand that $\overline{S}_n \cap \overline{T}_n = \emptyset$. However, since S_n and T_n are finite sets, it follows that $\overline{S}_n = S_n$, $\overline{T}_n = T_n$, and therefore $\overline{S}_n \cap \overline{T}_n = \emptyset$.

Remark 4.1. If s_{i_0} is $\mathcal{A}_{i_0}$-special, we shall consider in $S_n \cup T_n$ only this $\mathcal{A}_{i_0}$-special ultrafilter and no other $\mathcal{A}_{i_0}$-special ultrafilters, and no $\mathcal{A}_{i_0}$-similar ultrafilters. If s_{i_0}, t_{i_0} are $\mathcal{A}_{i_0}$-similar ultrafilters, we shall consider in $S_n \cup T_n$ only this pair of $\mathcal{A}_{i_0}$-ultrafilters and no other pairs of $\mathcal{A}_{i_0}$-similar ultrafilters, and no $\mathcal{A}_{i_0}$-special ultrafilters.

Let $s \in S_n$ and suppose there exists no pair s_i, t_i such that $s = s_i$. Then the singleton $\{s\}$ will be called a *bush of the first kind of* (S_n, T_n). Obviously, there exists s_i such that $s = s_i$ and s_i is a $\mathcal{A}_i$-special ultrafilter. Possibly there exists $j \neq i$ such that $s = s_j$, and s_j is a $\mathcal{A}_j$-special ultrafilter.

Let $t \in T_n$ and suppose there exist no pairs s_i, t_i and s_j, t_j such that $t = t_i \neq t_j$, $s_i = s_j$. The set

$$\{s_i \mid t_i = t\} \cup \{t\}$$

will be called a *bush of the second kind of* (S_n, T_n); $\{t\}$ will be called the *root* and $\{s_i \mid t_i = t\}$ the *crown* of the bush. A graphic representation of a bush of the second kind might be as follows:

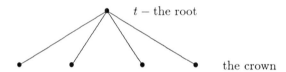

t − the root

the crown

If s belongs to the crown, then there must exist a pair s_i, t_i such that $s = s_i$, $t = t_i$. There may exist a pair s_j, t_j such that $s = s_j$, $t = t_j$ ($i \neq j$). There may exist an $\mathcal{A}_m$-special ultrafilter $s_m = s$. The crown of a bush of the second kind may contain only one point.

Let $s \in S$ be an ultrafilter contained neither in a bush of the first kind of (S_n, T_n) nor in a bush of the second kind of (S_n, T_n). Suppose there exist no pairs s_i, t_i and s_j, t_j such that $s = s_i \neq s_j$, $t_i = t_j$. The set

$$\{t_i \mid s_i = s\} \cup \{s\}$$

will be called a *bush of the third kind of* (S_n, T_n); $\{s\}$ will be called the *root* and $\{t_i \mid s_i = s\}$ the *crown* of the bush. A graphic representation of a bush of the third kind might be as follows:

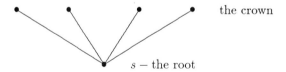

the crown

s − the root

If t belongs to the crown, then there must exist a pair s_i, t_i such that $s = s_i$, $t = t_i$. There may exist a pair s_j, t_j such that $s = s_j$, $t = t_j$ ($i \neq j$). There may exist an $\mathcal{A}_m$-special ultrafilter $s_m = s$. The crown of a bush of the third kind cannot contain only one point, since in that case it would be a bush of the second kind.

Let $t \in T_n$ and suppose that t is contained neither in a bush of the second kind of (S_n, T_n) nor in a bush of the third kind of (S_n, T_n). Let

$$P_1 = \{s_i \mid t_i = t\},$$
$$P_2 = \{t_i \neq t \mid s_i \in P_1\}.$$

It is clear that $P_2 \neq \emptyset$, since if $P_2 = \emptyset$, then t is the root of a bush of the second kind of (S_n, T_n). Next, define

$$P_3 = \{s_i \notin P_1 \mid t_i \in P_2\}.$$

Possibly $P_3 = \emptyset$. If $P_3 \neq \emptyset$, we continue the construction:

$$P_j = \{t_i \notin P_{j-2} \mid s_i \in P_{j-1}\},$$

if $j \geq 4$ is even;

$$P_j = \{s_i \notin P_{j-2} \mid t_i \in P_{j-1}\},$$

if $j \geq 5$ is odd. The construction ends when $P_j = \emptyset$. The set

$$\{t\} \cup \bigcup_j P_j$$

is called a *bush of the fourth kind of* (S_n, T_n). Here is a diagram of a simple bush of the fourth kind:

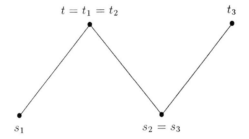

It is possible, for example, that $s_5 = s_1$, $t_5 = t$, and $s_{10} = s_2$, $t_{10} = t_3$. It may happen, for example, that s_{11} is an $\mathcal{A}_{11}$-special ultrafilter and $s_{11} = s_2$.

Clearly, $S_n \cup T_n$ is partitioned into bushes.

If K is a bush of (S_n, T_n), we define

$$\|K\| = |\{i,\ 1 \leq i \leq n \mid s_i \in K\}|.$$

It is easy to verify that

$$|K| \leq \|K\| + 1.$$

If K contains an $\mathcal{A}_i$-special ultrafilter s_i, then $|K| \leq ||K||$.

Let

$$n_0 = |\{i,\ 1 \leq i \leq n \mid s_i \text{ is contained in a bush of the fourth kind of } (S_n, T_n)\}|,$$

and suppose there are ℓ_0 bushes of the fourth kind in (S_n, T_n). It is clear that

$$n' = |\{q \in S_n \cup T_n \mid q \text{ is contained in a bush of the fourth kind of }$$
$$(S_n, T_n)\}| \leq n_0 + \ell_0.$$

If K is a bush of the fourth kind of (S_n, T_n), then $||K|| \geq 3$. Therefore,

$$n' \leq n_0 + \frac{n_0}{3} = \frac{4}{3} n_0.$$

Let

$$C = \{q \in S_n \cup T_n \mid q \text{ is the root of a bush of the second or third kind of }$$
$$(S_n, T_n) \text{ which contains no } \mathcal{A}_i\text{-special ultrafilters } s_i\}.$$

It is easy to verify that

$$|(S_n \cup T_n)\backslash C| \leq \frac{4}{3} n_0 + n - n_0 \leq \frac{4}{3} n.$$

One more fact will be needed. Let K be a bush that does not contain an $\mathcal{A}_i$-special ultrafilter s_i. Then K can be "inverted": if $s_i, t_i \in K$, denote s_i by t_i and t_i by s_i. Under this transformation, a bush of the second kind $K(|K| = 2)$ remains a bush of the second kind; a bush of the second kind $K(|K| > 2)$ becomes a bush of the third kind; a bush of the third kind becomes a bush of the second kind; a bush of the fourth kind remains a bush of the fourth kind. The new sets S_n, T_n thus obtained have the same properties as before, and if $Q \in s$ for all $s \in S_n$, but $Q \notin t$ for all $t \in T_n$, then $Q \notin \mathcal{A}_i$, $1 \leq i \leq n$.

Theorem 4.1. *Consider a finite sequence $\mathcal{A}_1, \ldots, \mathcal{A}_n, \mathcal{A}_{n+1}$ of algebras. Let $D \notin \mathcal{A}_k$, $1 \leq k \leq n$. Assume moreover that there are more than $\frac{4}{3} n$ pairwise disjoint sets not members of $\mathcal{A}_{n+1}$. Then there exists a set not a member of any of the algebras $\mathcal{A}_k$, $1 \leq k \leq n+1$.*

Proof. Consider sets S_n, T_n for the algebras $\mathcal{A}_1, \ldots, \mathcal{A}_n$. Our previous arguments and the fact that $|\ker \mathcal{A}_{n+1}| > \frac{4}{3}n$ imply that there exists $q \in \ker \mathcal{A}_{n+1}$ satisfying one of the following two conditions:

(1) $q \notin S_n \cup T_n$;

(2) q is contained in a bush K of (S_n, T_n); K is of either the second of the third kind; q is the root of K; K does not contain an $\mathcal{A}_i$-special ultrafilter $s_i \in S_n$, $1 \le i \le n$.

Case I. q is an $\mathcal{A}_{n+1}$-special ultrafilter. Let $q \in K$ and suppose that K is a bush of the second kind of $\{S_n, T_n\}$. Invert K, i.e., consider new sets S_n, T_n, with K becoming a bush K' of (S_n, T_n). If K' is of the second kind ($|K'| = 2$), then q is the crown of K'. If K' is of the third kind ($|K'| > 2$), then q is the root of K'. If $q \notin S_n \cup T_n$ or K is bush of the third kind of (S_n, T_n), leave S_n, T_n unchanged. Define $s_{n+1} = q$,

$$S_{n+1} = S_n \cup \{s_{n+1}\}, \quad T_{n+1} = T_n.$$

Note that after inversion, the $\mathcal{A}_{n+1}$-special ultrafilter s_{n+1} is contained in a bush either of the second kind or of the third kind of (S_{n+1}, T_{n+1}).

Case II. q, r are $\mathcal{A}_{n+1}$-similar ultrafilters, and q, r are not both in S_n or both in T_n. Then one of q, r, say r, is not in T_n. If $q \in S_n$, define $s_{n+1} = q$, $t_{n+1} = r$; if $q \notin S_n$, define $s_{n+1} = r$, $t_{n+1} = q$. Naturally,

$$S_{n+1} = S_n \cup \{s_{n+1}\}, \quad T_{n+1} = T_n \cup \{t_{n+1}\}.$$

Case III. Either $q, r \in S_n$, or $q, r \in T_n$. Suppose, say, that $q, r \in S_n$. Thus K is a bush of the third kind of (S_n, T_n). Invert K, i.e., consider new S_n, T_n, with K now a bush of the second kind K' of (S_n, T_n). When this is done, q becomes the root of K'. Define $s_{n+1} = r$, $t_{n+1} = q$,

$$S_{n+1} = S_n, \quad T_{n+1} = T_n.$$

Note that after inversion, q is contained in a bush either of the fourth kind or of the second kind of (S_{n+1}, T_{n+1}). In the latter case this bush contains an $\mathcal{A}_i$-special ultrafilter coinciding with r. If $q, r \in T_n$ the reasoning is analogous, with $s_{n+1} = q$, $t_{n+1} = r$. But in that case q can be contained only in a bush of the fourth kind of (S_{n+1}, T_{n+1}).

Thus, in all cases we have been able to construct sets S_{n+1}, T_{n+1} for the algebras $\mathcal{A}_1, \ldots, \mathcal{A}_n, \mathcal{A}_{n+1}$. $\square$

We have already pointed out (in the Introduction) that if $\mathcal{A}_1, \mathcal{A}_2 \neq \mathcal{P}(X)$, then there exists a set which is not a member of either of these algebras. Hence Theorem 4.1 implies the first part of Theorem I.

Remark 4.2. Consider a countable sequence $\mathcal{A}_1, \ldots, \mathcal{A}_k, \ldots$ of algebras, $\mathcal{A}_2 \neq \mathcal{P}(X)$, and suppose that for every $k \neq 2$ there exist more than $\frac{4}{3}(k-1)$ pairwise disjoint sets not members of $\mathcal{A}_k$. As in the proof of Theorem 4.1, construct sets

$$S_1, T_1; \ldots; S_k T_k; \ldots$$

inductively. Let $a \in S_k$ and $a \in T_{k+1}$, or $a \in T_k$ and $a \in S_{k+1}$. Then there are three possibilities: a is contained in a bush of the second or the third kind of (S_{k+1}, T_{k+1}) which contains an $\mathcal{A}_i$-special ultrafilter $s_i \in S_{k+1}$, $1 \leq i \leq k+1$; or a is contained in a bush of the fourth kind of (S_{k+1}, T_{k+1}). Therefore, if $a \in T_{k+1}$, then $a \in T_\ell$ for all $\ell > k$. If $a \in S_{k+1}$, then $a \in S_\ell$ for all $\ell > k$. It follows that there exist sets of ultrafilters S, T such that

$$S \cup T = \bigcup_k (S_k \cup T_k),$$
$$S \cap T = \emptyset,$$
$$S = \{s_1, \ldots, s_k, \ldots\},$$

and for every k one of the following conditions holds:

(1) s_k is an $\mathcal{A}_k$-special ultrafilter;

(2) there exist $\mathcal{A}_k$-similar ultrafilters s_k, t_k.

The set T is the collection of all t_k's. We cannot state, however, that $\overline{S} \cap \overline{T} = \emptyset$. Therefore, we cannot state that there exists a set not a member of any $\mathcal{A}_k$. In addition, such a set may not exist if it is not stipulated that the algebras $\mathcal{A}_k$ are almost σ-algebras (see Example 6.1).

The following theorem is a refined version of the first part of Theorem I:

Theorem 4.2. *Consider a finite sequence of algebras $\mathcal{A}_1, \ldots, \mathcal{A}_n$. Suppose that for every k, $1 \leq k \leq n$, we are given a finite sequence of pairwise disjoint sets $Q_1^k, \ldots, Q_{m_k}^k \notin \mathcal{A}_k$, where $m_k > \frac{4}{3}(k-1)$ if $k \neq 2$. Then for every set Q_i^k one can construct a set $\hat{Q}_i^k \subset Q_i^k$ such that $\hat{Q}_i^k \notin \mathcal{A}_k$; either $\hat{Q}_i^k = \hat{Q}_j^\ell$ or $\hat{Q}_i^k \cap \hat{Q}_j^\ell = \emptyset$; there exists a set Q which is the union of sets of the form $\hat{Q}_i^p$ such that $Q \notin \mathcal{A}_k$ for all k, $1 \leq k \leq n$.*

Proof. To each set Q_i^k we associate either an $\mathcal{A}_k$-special ultrafilter s_i^k or $\mathcal{A}_k$-similar ultrafilters s_i^k, t_i^k; $s_i^k \ni Q_i^k, t_i^k \not\ni Q_i^k$. Our goal is to construct suitable sets of ultrafilters S_n and T_n such that

$$S_n \subset \bigcup_{k,i} \{s_i^k\}.$$

Obviously, the possibility of constructing such sets S_n and T_n will imply the statement of our theorem. The construction proceeds by induction, i.e., we successively construct sets

$$S_1, T_1; \ldots; S_p, T_p; \ldots; S_n, T_n.$$

This will be done in such a way that

$$S_p \subset \bigcup_{k \leq p, i} \{s_i^k\}.$$

If $n = 1$, we define $S_1 = \{s_1^1\}$ and $T_1 = \{t_1^1\}$. If t_1^1 does not exist, then $T_1 = \emptyset$.

Let $n = 2$. If $t_1^2 = s_1^1$, then $S_2 = \{s_1^1\}$. If $t_1^2 \neq s_1^1$ and $t_1^1 = s_1^2$, then $S_2 = \{s_1^1\}$. In all other cases $S_2 = \{s_1^1, s_1^2\}$. The construction of T_2 is quite natural.

Let $1 < p < n$ and suppose that the sets S_p, T_p have already been constructed. We proceed to construct S_{p+1}, T_{p+1}. As in the proof of Theorem 4.1, consider an ultrafilter

$$q \in \{s_1^{p+1}, \ldots, s_{m_{p+1}}^{p+1}\} = S_*$$

satisfying one of the following two conditions:

(1) $q \notin S_p \cup T_p$;

(2) q is contained in a bush K of (S_p, T_p); K is of either the second or the third kind; q is the root of K; K does not contain an $\mathcal{A}_i$-special ultrafilter $s_i \in S_p$, $1 \leq i \leq p$.

In addition, we may assume that q satisfies the condition

(3) if q satisfies condition (2), then

$$K \subset S_*.^{13}$$

Indeed, suppose that among all bushes of the second and third kind (S_p, T_p), which do not contain the $\mathcal{A}_i$-special ultrafilters s_i, $1 \leq i \leq p$, there are τ bushes whose crowns are not contained in S_*. Then there exist more than τ ultrafilters in S_*, each of which satisfies either condition (1) or (2). Therefore, there exists q satisfying either condition (1) or conditions (2) and (3). If q is an $\mathcal{A}_{p+1}$-special ultrafilter, we reason exactly as in the proof of Theorem 4.1. Otherwise, as in the proof of Theorem 4.1, we consider $\mathcal{A}_{p+1}$-similar ultrafilters q, r. If $r \in S_p \cup T_p$, we reason as in the proof of Theorem 4.1. If $r \notin S_p \cup T_p$, we put

$$s_{p+1} = q, \quad t_{p+1} = r$$

and if necessary invert the bush K. In all cases

$$S_{p+1} \subset S_p \cup S_*. \quad \square$$

Remark 4.3. Suppose that the sequence of algebras in Theorem 4.2 is not finite but countable, $\mathcal{A}_1, \ldots, \mathcal{A}_k, \ldots$. Then an analysis of the proof of the theorem will show that we can construct sets S, T as in Remark 4.2 such that also

$$S \subset \bigcup_{k,i} \{s_i^k\}.$$

In what follows, we shall not use this remark.

Lemma 4.1. *Let $\mathcal{A}_1, \ldots, \mathcal{A}_n, \mathcal{A}_{n+1}$ be a finite sequence of algebras. There exists a set not belonging to all algebras $\mathcal{A}_k$, $1 \leq k \leq n$; S_n and T_n are corresponding sets of ultrafilters; there are more than $\frac{4}{3}n$ pairwise disjoint sets not in $\mathcal{A}_{n+1}$. Suppose also that we have a bush K of (S_n, T_n), containing r points. Then there exist pairwise distinct $q_1, \ldots, q_p \in \ker \mathcal{A}_{n+1}$*

---

[13]It is thanks to condition (3), which was not assumed in the proof of Theorem 4.1, that we can improve the first part of Theorem I.

such that $p > \frac{r}{3} - \alpha$ (if K is a bush of the fourth kind $\alpha = \frac{4}{3}$; otherwise $\alpha = \frac{1}{3}$), and for every j, $1 \leq j \leq p$, one of the following two conditions holds:

(1) $q_j \notin S_n \cup T_n$;

(2) q_j is contained in a bush K_j of (S_n, T_n); K_j is a bush of either the second or third kind; q_j is the root of K_j; K_j does not contain an $\mathcal{A}_i$-special ultrafilter $s_i \in S_n$, $1 \leq i \leq n$.

Proof. We shall use the following formula which has already been established:

$$|(S_n \cup T_n)\backslash C| \leq n' + (n - n_0) \leq \frac{4}{3}n_0 + (n - n_0).$$

Obviously,

$$p > \frac{4}{3}n - |(S_n \cup T_n)\backslash C| \geq \frac{n - n_0}{3}.$$

We consider three cases:

Case I. If K is a bush of the first kind, then $r = 1$, $n - n_0 \geq 1$ and $p > 0$.

Case II. If K is a bush of the second or third kind, then $n - n_0 \geq r - 1$ and therefore

$$p > \frac{r}{3} - \frac{1}{3}.$$

Case III. If K is a bush of the fourth kind, then relying on the previous arguments, we have

$$|\{q \in S_n \cup T_n \mid q \text{ is contained in a bush of the fourth kind of}$$
$$(S_n, T_n), \text{ and } q \notin K\}| \leq \frac{4}{3}(n_0 - ||K||).$$

Thus,

$$n' \leq r + \frac{4}{3}(n_0 - ||K||).$$

Since $||K|| \geq r - 1$, it follows that

$$|(S_n \cup T_n)\backslash C| \leq r + \frac{4}{3}(n_0 - r + 1) + (n - n_0).$$

Therefore,

$$p > \frac{4}{3}n - [r + \frac{4}{3}(n_0 - r + 1) + (n - n_0)] = \frac{n - n_0}{3} + \frac{r}{3} - \frac{4}{3}. \quad \Box$$

Remark 4.4. Lemma 4.1 will be used in the proof of Theorem II. Essentially, however, the only important point for the proof of the latter theorem is: if $|K|$ is a "large number", then p is also "large".

We now proceed to the

Proof of the second part of Theorem I. Consider pairwise distinct points in βX :

$$\begin{pmatrix} a_1^1 & a_2^1 & b_1^1 & b_2^1 \\ a_1^2 & a_2^2 & b_1^2 & b_2^2 \\ \cdots\cdots\cdots\cdots\cdots\cdots \\ a_1^n & a_2^n & b_1^n & b_2^n \end{pmatrix}.$$

Define algebras $\mathcal{A}_1, \ldots, \mathcal{A}_{3n}$ as follows:[14]

$$\ker \mathcal{A}_1 = \{a_1^1, b_1^1\};$$

$$\ker \mathcal{A}_2 = \{a_2^1, b_1^1\};$$

$$\ker \mathcal{A}_3 = \{a_1^1, a_2^1, b_2^1\}; \quad [15]$$

$$\ker \mathcal{A}_4 = \{a_1^1, a_2^1, b_1^1, b_2^1, a_1^2, b_1^2\},$$

and $\{a_1^1, a_2^1\}$, $\{b_1^1, b_2^1\}$, $\{a_1^2, b_1^2\}$ are $\mathcal{A}_4$-similar sets;

$$\ker \mathcal{A}_5 = \{a_1^1, a_2^1, b_1^1, b_2^1, a_2^2, b_1^2\},$$

and $\{a_1^1, a_2^1\}$, $\{b_1^1, b_2^1\}$, $\{a_2^2, b_1^2\}$ are $\mathcal{A}_5$-similar sets;

$$\ker \mathcal{A}_6 = \{a_1^1, a_2^1, b_2^1, b_2^1, a_1^2, a_2^2, b_2^2\},$$

and $\{a_1^1, a_2^1\}$, $\{b_1^1, b_2^1\}$, $\{a_1^2, a_2^2, b_2^2\}$ are $\mathcal{A}_6$-similar sets.

---

[14] For every algebra $\mathcal{A}$ included in this construction, $X \in \mathcal{A}$; hence there exist no $\mathcal{A}$-special ultrafilters.

[15] Since $X \in \mathcal{A}_1, \mathcal{A}_2, \mathcal{A}_3$, it follows that $\{a_1^1, b_1^1\}$ is a $\mathcal{A}_1$-similar set, $\{a_2^1, b_1^1\}$ is a $\mathcal{A}_2$-similar set, and $\{a_1^1, a_2^1, b_2^1\}$ is a $\mathcal{A}_3$-similar set.

In general: The $\mathcal{A}_{3n-2}$-similar sets are $\{a_1^i, a_2^i\}$, $\{b_1^i, b_2^i\}$, $1 \le i \le n-1$, and also $\{a_1^n, b_1^n\}$. The algebra $\mathcal{A}_{3n-1}$ differs from $\mathcal{A}_{3n-2}$ only in that $\{a_1^n, b_1^n\}$ is replaced by $\{a_2^n, b_1^n\}$ as an $\mathcal{A}_{3n-1}$-similar set. Finally, $\mathcal{A}_{3n}$ differs from $\mathcal{A}_{3n-2}$ only in that $\{a_1^n, b_1^n\}$ is replaced by $\{a_1^n, a_2^n, b_2^n\}$ as an $\mathcal{A}_{3n}$-similar set.

Obviously, there are more than $\frac{4}{3}(k-1)$ pairwise disjoint sets not members of $\mathcal{A}_k$, $1 \le k \le 3n$.

It is convenient to represent the points a_j^i and b_j^i by a diagram:

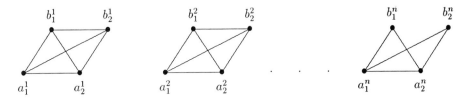

Consider the algebra $\mathcal{A}_{3n+1}^*$ defined as follows. The $\mathcal{A}_{3n+1}^*$-similar sets are $\{a_1^i, a_2^i\}$, $\{b_1^i, b_2^i\}$, $1 \le i \le n$. It is readily verified that any subset of X is a member of at least one of the algebras $\mathcal{A}_1, \ldots, \mathcal{A}_{3n}, \mathcal{A}_{3n+1}^*$. Indeed, if $Q \notin \mathcal{A}_k$, $1 \le k \le 3n$, then $\overline{Q}$ must contain one of the following two pairs:

$$\text{either } a_1^1, a_2^1 \text{ or } b_1^1, b_2^1.$$

Then $\overline{Q}$ must contain one of the following two pairs:

$$\text{either } a_1^2, a_2^2 \text{ or } b_1^2, b_2^2,$$

and so on. Then $\overline{Q}$ must contain one of the following two pairs:

$$\text{either } a_1^n, a_2^n \text{ or } b_1^n, b_2^n.$$

But such $Q \in \mathcal{A}_{3n+1}^*$.

Clearly, there exist $4n$ pairwise disjoint sets not members of $\mathcal{A}_{3n+1}^*$. If there were more than $4n$ such sets, it would follows from the first part of Theorem I that there exists a set not a member of any of the algebras $\mathcal{A}_1, \ldots, \mathcal{A}_{3n}, \mathcal{A}_{3n+1}^*$.

Now consider the algebras $\mathcal{A}_{3n+1}'$, $\mathcal{A}_{3n+2}'$ defined as follows. Let $q \in \beta X$, and q is not one of the points a_j^i, b_j^i. All $\mathcal{A}_{3n+1}^*$-similar sets except for $\{a_1^n, a_2^n\}$ are $\mathcal{A}_{3n+1}'$-similar. There exists one more $\mathcal{A}_{3n+1}'$-similar set, $\{a_1^n, a_2^n, q\}$. All $\mathcal{A}_{3n+1}^*$-similar sets except for $\{b_1^n, b_2^n\}$ are $\mathcal{A}_{3n+2}'$-similar. There exists one more $\mathcal{A}_{3n+2}'$-similar set, $\{b_1^n, b_2^n, q\}$. It is readily verified that any subset of X is a member of at least one of the algebras $\mathcal{A}_1, \ldots, \mathcal{A}_{3n}, \mathcal{A}_{3n+1}', \mathcal{A}_{3n+2}'$.

Indeed, let $Q \notin \mathcal{A}_k$ $(1 \le k \le 3n)$, $Q \notin \mathcal{A}'_{3n+1}, \mathcal{A}'_{3n+2}$. Then either $a_1^i, a_2^i \subset \overline{Q}$ or $b_1^i, b_2^i \subset \overline{Q}$. Let $a_1^n, a_2^n \subset \overline{Q}$. If $q \in Q$, then $Q \in \mathcal{A}'_{3n+1}$. But if $q \notin \overline{Q}$, then $Q \in \mathcal{A}'_{3n+2}$.

Clearly, $|\ker \mathcal{A}'_{3n+1}| = 4n+1$. Obviously, there exist $4n+1$ pairwise disjoint sets not members of $\mathcal{A}'_{3n+2}$, for if there were more, the first part of Theorem I would imply the existence of a set not a member of any of the algebras $\mathcal{A}_1, \dots, \mathcal{A}_{3n}, \mathcal{A}'_{3n+1}, \mathcal{A}'_{3n+2}$.

Now consider the algebras $\hat{\mathcal{A}}_{3n+1}, \hat{\mathcal{A}}_{3n+2}, \hat{\mathcal{A}}_{3n+3}$ defined as follows. Let q_1, q_2, q_3 be distinct points in βX. The points q_m are distinct from any of a_j^i, b_j^i. All $\mathcal{A}^*_{3n+1}$-similar sets are $\hat{\mathcal{A}}_{3n+i}$-similar, $i = 1, 2, 3$. For each $\hat{\mathcal{A}}_{3n+i}$ there is one more $\hat{\mathcal{A}}_{3n+i}$-similar set: for $\hat{\mathcal{A}}_{3n+1}$ $- \{q_1, q_2\}$; for $\hat{\mathcal{A}}_{3n+2} - \{q_2, q_3\}$; and for $\hat{\mathcal{A}}_{3n+3} - \{q_1, q_3\}$. It is readily verified that any subset of X is a member of at least one of the algebras $\mathcal{A}_1, \dots, \mathcal{A}_{3n}, \hat{\mathcal{A}}_{3n+1}, \hat{\mathcal{A}}_{3n+2}, \hat{\mathcal{A}}_{3n+3}$. Indeed, let $Q \notin \mathcal{A}_k$ $(q \le k \le 3n)$, $Q \notin \hat{\mathcal{A}}_{3k+i}$ $(i = 1, 2, 3)$. If $q_1, q_2, q_3 \notin \overline{Q}$ or $q_1, q_2, q_3 \in \overline{Q}$, then $Q \in \hat{\mathcal{A}}_{3n+i}$ $(i = 1, 2, 3)$. But if, say, $q_1 \in \overline{Q}$ and $q_2, q_3 \notin \overline{Q}$, then $Q \in \hat{\mathcal{A}}_{3n+2}$.

Clearly,

$$|\ker \hat{\mathcal{A}}_{3n+1}| = |\ker \hat{\mathcal{A}}_{3n+2}| = 4n+2.$$

Obviously, there exist $4n+2$ pairwise disjoint sets not members of $\hat{\mathcal{A}}_{3n+3}$, for if there were more, the first part of Theorem I implies the existence of a set not a member of any of the algebras $\mathcal{A}_1, \dots, \mathcal{A}_{3n}, \hat{\mathcal{A}}_{3n+1}, \hat{\mathcal{A}}_{3n+2}, \hat{\mathcal{A}}_{3n+3}$.

Our constructions imply the statement of the second part of Theorem I. $\square$

Definition 4.1. If there exists a two-valued measure such that every subset of X is measurable, then $|X|$ is called a σ-measurable cardinal. A cardinal which is not σ-measurable is said to be σ-nonmeasurable.[16]

The assumption that all cardinals are σ-nonmeasurable is known to be consistent with the axioms of set theory.

---

[16]Some authors (see, e.g. [E]) use the term "measurable" for what we have called a σ-measurable cardinal; however, the standard definition of a measurable cardinal is different: an infinite cardinal k is said to be measurable if one can define on X, $|X| = k$, a k-additive measure μ such that $\mu(\{x\}) = 0$, $x \in X$, and $\mu(X) = 1$. In addition, if M is a μ-measurable set, then either $\mu(M) = 0$ or $\mu(M) = 1$. (A measure μ is said to be k-additive if, for any family of μ-measurable sets $\{M_\lambda\}_{\lambda \in \Lambda}$, $|\Lambda| < k$, the set $\bigcup_\lambda M_\lambda$ is μ-measurable; and if $\{M_\lambda\}_{\lambda \in \Lambda}$ is a family of pairwise disjoint sets, then $\mu(\bigcup_\lambda M_\lambda) = \sum_\lambda \mu(M_\lambda)$.) Therefore, in order to avoid confusion, we have introduced the concept of a σ-measurable cardinal. This definition seems to us to be logical, since it is concerned with a two-valued measure, which is σ-additive. Let k_0 be the smallest uncountable measurable cardinal. It is known that a cardinal k is σ-measurable if and only if $k \ge k_0$.

Definition 4.2. A point $a \in \beta X$ is said to be irregular if there is no countable sequence $M_1, \ldots, M_k, \ldots \subset X$ such that $a \notin \overline{M}_k$ for all k, but $a \in \overline{\bigcup_k M_k}$. A point of βX is said to be regular if it is not irregular.

If $a \in X$, then a is an irregular point. A point a is regular if and only if there exists a real-valued function f, continuous on βX, such that $f(a) = 0$ and $f(x) > 0$ for $x \in X$. If $|X|$ is a σ-measurable cardinal and μ a two-valued measure defined on all subsets of X, then the collection of sets $M \subset X$ such that $\mu(M) = 1$ is an ultrafilter $a \in \beta X \backslash X$ and a is an irregular point. On the other hand, a point $a \in \beta X \backslash X$ is irregular if and only if, putting $\mu(M) = 1$ for $M \in a$, one obtains a two-valued measure defined on all subsets of X. Thus, the cardinal $|X|$ is σ-measurable if and only if $\beta X \backslash X$ contains irregular points.

Let us assume that $|X|$ is a σ-measurable cardinal and the points $a_j^i, b_j^i, q, q_1, q_2, q_3$, which were considered in the proof of the second part of Theorem I, are irregular points in $\beta X \backslash X$. Then any algebra $\mathcal{A}$ which was considered in the proof of the second part of Theorem I is an algebra of ν-measurable sets, while ν is a σ-additive measure and $\nu(X) = 1$, $\nu(\{x\}) = 0$ for $x \in X$.[17] Obviously, there exist at most finitely many pairwise disjoint ν-nonmeasurable sets. A measure ν satisfying these conditions cannot exist if we assume that $|X|$ is a σ-nonmeasurable cardinal. This is a corollary of the following simple proposition:

Proposition. *If $|X|$ is a σ-nonmeasurable cardinal, $\mathcal{A}$ is an almost σ-algebra, there exists a set not in $\mathcal{A}$, and $\{x\} \in \mathcal{A}$ for all $x \in X$, then there exist $\aleph_0$ pairwise disjoint sets not in $\mathcal{A}$.*

5. Countable Sequences of Algebras (1)

Definition 5.1. An algebra $\mathcal{A}$ is said to be simple if there exists a set $Z \subset \beta X$ such that

(1) $|Z| \leq \aleph_0$;

(2) if $z \in Z$, then z is an irregular point;

(3) $\ker \mathcal{A} \subset \overline{Z}$.

---

[17] It is very easy to prove the existence of the measure ν (see Remark 6.3).

The aim of this section is to prove the crucially important Theorem 5.3: thanks to this theorem, the main object of this memoir can be pursued by considering only simple algebras.

In the sequel we shall use the following interesting theorem:

Theorem 5.1. *Consider a countable sequence of algebras* $\mathcal{A}_1, \ldots, \mathcal{A}_k, \ldots$. *Assume that there is a matrix*

$$\begin{pmatrix} U_1^1 & & \\ U_1^2 & U_2^2 & \\ \cdots\cdots\cdots & \\ U_1^k & \cdots & U_k^k \\ \cdots\cdots\cdots & \end{pmatrix}$$

of pairwise disjoint sets such that $U_i^k \notin \mathcal{A}_k$. *Let* $W \subset X \setminus \bigcup_{k,i} U_i^k$. *Then there exists a set* $W \cup U \notin \mathcal{A}_k$ *for all* k. *In addition, either* $U = \emptyset$ *or* $U = \bigcup_i U_{\beta_i}^{\alpha_i}$, *where* $\alpha_1 < \alpha_2 < \ldots < \alpha_i < \ldots$ *is a finite or an infinite sequence.*[18]

Proof. From two sets $W, W \cup U_1^1$ choose one not belonging to $\mathcal{A}_1$, and define this set by V_1. If we assume that this set does not exist, then

$$(W \cup U_1^1) \setminus W = U_1^1 \in \mathcal{A}_1.$$

But this is false.

Assume that there exists a finite or infinite sequence

$$U_{\beta_1^1}^{\alpha_1^1}, U_{\beta_2^1}^{\alpha_2^1}, \ldots, U_{\beta_m^1}^{\alpha_m^1}, \ldots \quad (1 < \alpha_1^1 < \alpha_2^1 < \ldots < \alpha_m^1 < \ldots)$$

such that

$$\mathcal{A}_1 \ni V_1 \cup \bigcup_m U_{\beta_m^1}^{\alpha_m^1}.$$

In this case all the sets

$$U_{\beta_1^1}^{\alpha_1^1}, U_{\beta_2^1}^{\alpha_2^1}, \ldots, U_{\beta_m^1}^{\alpha_m^1}, \ldots$$

are called distinguished.

---

[18] By a slight modification of the proof of the first part of Theorem IV, one can prove Theorem 5.1 – instead of the matrix appearing in Theorem 5.1, one should take the matrix in Theorem IV. That is to say, Theorem 5.1 is a special case of the first part of Theorem IV. The proof of the first part of Theorem IV is rather complicated. On the other hand, Theorem 5.1, which is easier to prove, is quite sufficient for the proof of Theorem 5.3 and other theorems.

Now consider the sets U_1^2, U_2^2. At least one of them, say U_2^2, is not distinguished. ¿From two sets $V_1, V_1 \cup U_2^2$ choose one which does not belong to $\mathcal{A}_2$, and define this set by V_2. Assume that there exists a finite or infinite sequence

$$U_{\beta_1^2}^{\alpha_1^2}, U_{\beta_2^2}^{\alpha_2^2}, \ldots, U_{\beta_m^2}^{\alpha_m^2}, \ldots \quad (2 < \alpha_1^2 < \alpha_2^2 < \ldots < \alpha_m^2 < \ldots)$$

such that

$$\mathcal{A}_2 \ni V_2 \cup \bigcup_m U_{\beta_m^2}^{\alpha_m^2}.$$

In this case all the sets

$$U_{\beta_1^2}^{\alpha_1^2}, U_{\beta_2^2}^{\alpha_2^2}, \ldots, U_{\beta_m^2}^{\alpha_m^2}, \ldots$$

are called distinguished.

Continuing the construction, we consider a corresponding set V_{k-1} and the sets $U_1^k, \ldots, U_k^k$. There exists at least one undistinguished set U_j^k. We know that $V_{k-1} \cap U_j^k = \emptyset$. Therefore, from two sets $V_{k-1}, V_{k-1} \cup U_j^k$ it is possible to choose one which does not belong to $\mathcal{A}_k$. This set is defined by V_k. Assume that there exists a finite or infinite sequence

$$U_{\beta_1^k}^{\alpha_1^k}, U_{\beta_2^k}^{\alpha_2^k}, \ldots, U_{\beta_m^k}^{\alpha_m^k}, \ldots \quad (k < \alpha_1^k < \alpha_2^k < \ldots < \alpha_m^k < \ldots)$$

such that

$$\mathcal{A}_k \ni V_k \cup \bigcup_m U_{\beta_m^k}^{\alpha_m^k}.$$

In this case all the sets

$$U_{\beta_1^k}^{\alpha_1^k}, U_{\beta_2^k}^{\alpha_2^k}, \ldots, U_{\beta_m^k}^{\alpha_m^k}, \ldots$$

are called distinguished.

Consider all rows of the matrix of sets specified in the theorem, and construct the sequence of sets

$$V_1 \subset V_2 \subset \ldots \subset V_k \subset \ldots .$$

Let

$$U = (\bigcup_k V_k) \backslash W.$$

Obviously, either $U = \emptyset$ or $U = \bigcup_i U^{\alpha_i}_{\beta_i}$, where $\alpha_1 < \alpha_2 < \ldots < \alpha_i < \ldots$ is a finite or infinite sequence.

Prove that $W \cup U \notin \mathcal{A}_k$. This is certainly so if there is no sequence

$$U^{\alpha^k_1}_{\beta^k_1}, U^{\alpha^k_2}_{\beta^k_2}, \ldots, U^{\alpha^k_m}_{\beta^k_m}, \ldots .$$

If such a sequence exists and we assume that $W \cup U \in \mathcal{A}_k$, then

$$(W \cup U) \cap (V_k \cup \bigcup_m U^{\alpha^k_m}_{\beta^k_m}) = V_k \in \mathcal{A}_k.$$

But $V_k \notin \mathcal{A}_k$, a contradiction. $\square$

We shall not use the following proposition for proving our main theorems. The interest of this proposition is that it is proved with the help of a very easy technique which was used in the proof of Theorem 5.1, and that a generalization of this proposition is contained in Section 13.

Proposition 5.1. *Consider a countable sequence of algebras $\mathcal{A}_1, \ldots, \mathcal{A}_k, \ldots$. Assume that there is a matrix*

$$\begin{pmatrix} U^1_1 & & \\ U^2_1 & U^2_2 & U^2_3 \\ \cdots\cdots\cdots\cdots\cdots\cdots\cdots \\ U^k_1 \cdots\cdots\cdots & U^k_k & U^k_{k+1} \\ \cdots\cdots\cdots\cdots\cdots\cdots \end{pmatrix}$$

of pairwise disjoint sets such that $U^k_i \notin \mathcal{A}_k$. (The first row contains one set, and the k-th row ($k > 1$) contains $k + 1$ sets.) Then there exists a set $U = \bigcup_k U^k_{i_k}$ not belonging to all $\mathcal{A}_k$.

Proof. Assume that there exists a sequence

$$U^2_{i^1_2}, U^3_{i^1_3}, \ldots, U^j_{i^1_j}, \ldots$$

such that

$$\mathcal{A}_1 \ni U^1_1 \cup \bigcup_{j>1} U^j_{i^1_j}.$$

In this case all the sets $U^j_{i^1_j}$ ($j > 1$) are called distinguished.

Now consider the sets U_1^2, U_2^2, U_3^2. Among these sets there are two, say U_2^2, U_3^2, undistinguished sets. ¿From two sets $U_1^1 \cup U_2^2, U_1^1 \cup U_3^2$ choose one, say $U_1^1 \cup U_2^2$, not belonging to $\mathcal{A}_2$. This is possible because if $U_1^1 \in \mathcal{A}_2$, then $U_1^1 \cup U_2^2 \notin \mathcal{A}_2$. But if $U_1^1 \notin \mathcal{A}_2$, then

$$(U_1^1 \cup U_2^2) \cap (U_1^1 \cup U_3^2) \notin \mathcal{A}_2.$$

The sets U_1^2, U_3^2 are called distinguished.

Assume that there exists a sequence

$$U_{i_3^2}^3, U_{i_4^2}^4, \ldots, U_{i_j^2}^j, \ldots$$

such that

$$\mathcal{A}_2 \ni U_1^1 \cup U_2^2 \cup \bigcup_{j>2} U_{i_j^2}^j.$$

In this case all the sets $U_{i_2^2}^j$ $(j > 2)$ are called distinguished.

Now consider the sets $U_1^3, U_2^3, U_3^3, U_4^3$. Among these sets there are two, say U_1^3, U_2^3, undistinguished sets. ¿From two sets $U_1^1 \cup U_2^2 \cup U_1^3$, $U_1^1 \cup U_2^2 \cup U_2^3$ chooose one, say $U_1^1 \cup U_2^2 \cup U_1^3$, not belonging to $\mathcal{A}_3$. This is possible because if $U_1^1 \cup U_2^2 \in \mathcal{A}_3$, then $U_1^1 \cup U_2^2 \cup U_1^3 \notin \mathcal{A}_3$. But if $U_1^1 \cup U_2^2 \notin \mathcal{A}_3$, then

$$(U_1^1 \cup U_2^2 \cup U_1^3) \cap (U_1^1 \cup U_2^2 \cup U_2^3) \notin \mathcal{A}_3.$$

The sets U_2^3, U_3^3, U_4^3 are called distinguished, and so on.

Continuing the construction we shall have in each row of the matrix of sets specified in the theorem only one undistinguished set. The union of all undistinguished sets is U. $\square$

In the sequel we shall need the following

Lemma 5.1. *Let*

$$\begin{pmatrix} U_1^1 & \cdots & U_{n_1}^1 \\ \cdots\cdots\cdots\cdots \\ U_1^k & \cdots & U_{n_k}^k \\ \cdots\cdots\cdots\cdots \end{pmatrix}$$

be a matrix of pairwise disjoint subsets of X, $n_k \to \infty$, $Z \subset \beta X$, $|Z| \leq \aleph_0$ and $Z \cap \overline{U_i^k} = \emptyset$. Then there exists a sequence $U_{i_1}^1, U_{i_2}^2, \ldots, U_{i_k}^k, \ldots$ such that $\overline{Z} \cap \overline{\bigcup_k U_{i_k}^k} = \emptyset$.

Proof. Let

$$Z = \{z_1, \ldots, z_m, \ldots\}.$$

For each point z_m there is either some corresponding sequence of sets or there is none. Suppose there exists a sequence

$$U^{\ell_1^1}_{r_1^1}, U^{\ell_2^1}_{r_2^1}, \ldots, U^{\ell_j^1}_{r_j^1}, \ldots, \quad \ell_1^1 < \ell_2^1 < \ldots < \ell_j^1 < \ldots,$$

such that $z_1 \ni \bigcup_j U^{\ell_j^1}_{r_j^1}$, and there are at least two sets in row ℓ_j^1 of the matrix. Then we call all the sets $U^{\ell_j^1}_{r_j^1}$ distinguished sets. Suppose we have a sequence

$$U^{\ell_1^2}_{r_1^2}, U^{\ell_2^2}_{r_2^2}, \ldots, U^{\ell_j^2}_{r_j^2}, \ldots, \quad \ell_1^2 < \ell_2^2 < \ldots < \ell_j^2 < \ldots,$$

such that $z_2 \ni \bigcup_j U^{\ell_j^2}_{r_j^2}$, and there are at least two undistinguished sets in row ℓ_j^2 of the matrix. Then we call all the sets $U^{\ell_j^2}_{r_j^2}$ distinguished sets. Continuing the construction, we look for a sequence

$$U^{\ell_1^m}_{r_1^m}, U^{\ell_2^m}_{r_2^m}, \ldots, U^{\ell_j^m}_{r_j^m}, \ldots, \quad \ell_1^m < \ell_2^m < \ldots < \ell_j^m < \ldots,$$

such that $z_m \ni \bigcup_j U^{\ell_j^m}_{r_j^m}$, and there are at least two undistinguished sets in row ℓ_j^m of the matrix. Then we call all the sets $U^{\ell_j^m}_{r_j^m}$ distinguished sets, and so on. When the procedure has been carried out for all points of Z, we observe that the k-th row of the matrix of sets contains an undistinguished set $U^k_{i_k}$. Since X is discrete, it follows that

$$\overline{Z} \cap \overline{\cup_k U^k_{i_k}} = \emptyset. \;\; \square$$

Remark 5.1. Setting $n_k = k+1$ in the assumptions of Lemma 5.1, we can assume that $\ell_j^m = m + j - 1$. In subsequent applications of Lemma 5.1 no restrictions will be imposed on the choice of the numbers n_k. We shall therefore assume from now on that $n_k = k+1$.

The next step of our investigation will include the following three definitions and Theorem 5.2.

Definition 5.2. A subset C of a topological space Y is said to be outer separable if there exists an at most countable set $Z \subset Y$ such that $\overline{Z} \supset C$. A set which is not outer separable is said to be outer inseparable.

In the sequel, unless otherwise stated, we shall confine our attention to outer separable and outer inseparable subsets of the topological space βX.

Definition 5.3. Let C be a subset of a topological space Y and suppose there exists a point $y_0 \in Y$ such that for any neighborhood $U(y_0)$ the set $U(y_0) \cap C$ is outer inseparable. Then y_0 is called an inseparability point of C.

If Y is compact, then an inseparability point of C exists if and only if C is an outer inseparable set.

Definition 5.4. An algebra $\mathcal{A}$ is said to be separable if $\ker \mathcal{A}$ is an outer separable subset. An algebra which is not separable is said to be inseparable.

The following assertion is obvious:

Assertion. *An algebra $\mathcal{A}$ is separable if and only if there exists an at most countable family of ultrafilters $\{a_\lambda\}$ such that if $M \notin a_\lambda$ for all λ, then $M \in \mathcal{A}$.*

Theorem 5.2. *Consider a countable sequence $\mathcal{A}, \dots, \mathcal{A}_k, \dots$ of algebras. Suppose there exists a set Q such that if $\mathcal{A}_k$ is separable, then $Q \notin \mathcal{A}_k$. Then there exists a set that is not a member of any of these algebras.*

Proof. Let $\mathcal{A}_{k_1}, \dots, \mathcal{A}_{k_n}, \dots$ be the sequence of all inseparable algebras among $\mathcal{A}_1, \dots, \mathcal{A}_k$, $\dots$. Let q_n be an inseparability point of $\ker \mathcal{A}_{k_n}$, $Z \subset \beta X$, $|Z| \leq \aleph_0$, $q_n \in Z$, and $\overline{Z} \supset \ker \mathcal{A}_k$ if $\mathcal{A}_k$ is a separable algebra.

Since $\ker \mathcal{A}_{k_1}$ is outer inseparable, there exist disjoint sets $U_1^1, U_2^1 \subset X$ such that

$$Z \cap \overline{U_i^1} = \emptyset, \ U_i^1 \notin \mathcal{A}_{k_1}.$$

Let $V_2 \subset X$, $V_2 \in q_2$, $(U_1^1 \cup U_2^1) \cap V_2 = \emptyset$. Since $\overline{V}_2 \cap \ker \mathcal{A}_{k_2}$ is outer inseparable, there exist pairwise disjoint sets

$$U_1^2, \dots, U_6^2 \subset V_2$$

such that

$$Z \cap \overline{U_i^2} = \emptyset, \quad U_i^2 \notin \mathcal{A}_{k_2}.$$

Continuing the procedure, we construct a matrix of pairwise disjoint sets

$$\begin{pmatrix} U_1^1 & U_2^1 & \\ \hdotsfor{2} \\ U_1^n & \cdots & U_{n(n+1)}^n \\ \hdotsfor{2} \end{pmatrix}$$

such that $Z \cap \overline{U_i^n} = \emptyset$ and $U_i^n \notin \mathcal{A}_{k_n}$. Let

$$W_i^n = \bigcup_{m=P_{i,n}}^{ni} U_m^n,$$

$P_{i,n} = n(i-1)+1$, $1 \le i \le n+1$. We constructed a matrix of pairwise disjoint sets

$$\begin{pmatrix} W_1^1 & W_2^1 & \\ \hdotsfor{2} \\ W_1^n & \cdots & W_{n+1}^n \\ \hdotsfor{2} \end{pmatrix}$$

with $Z \cap \overline{W_i^n} = \emptyset$. By Lemma 5.1, there exists a sequence

$$W_{i_1}^1, \ldots, W_{i_n}^n, \ldots$$

such that

$$\overline{Z} \cap \overline{\bigcup_n W_{i_n}^n} = \emptyset.$$

Let $W^* = X \setminus \bigcup_n W_{i_n}^n$, $W = W^* \cap Q$ (cf. assumptions of the theorem). If $\mathcal{A}_k$ is a separable algebra, then $\ker \mathcal{A}_k \subset \overline{W^*}$ ($\ker \mathcal{A}_k \subset \overline{Z} \subset \overline{W^*}$), and therefore $W \notin \mathcal{A}_k$. Since $W_{i_n}^n$ contains n pairwise disjoint sets

$$U_{P_{i_n,n}}^n, \ldots, U_{ni_n}^n$$

and none of these sets is a member of $\mathcal{A}_{k_n}$, it follows from Theorem 5.1 that there exists a set $W \cup U \notin \mathcal{A}_{k_n}$ for all n. In addition, either $U = \emptyset$ or U is a union of sets $U_{j_n}^n$, $P_{i_n,n} \le j_n \le ni_n$. If $\mathcal{A}_k$ is a separable algebra, then $U \in \mathcal{A}_k$ ($Z \cap \overline{U} = \emptyset$, $\ker \mathcal{A}_k \subset \overline{Z}$), $W \notin \mathcal{A}_k$, $W \cap U = \emptyset$, and therefore $W \cup U \notin \mathcal{A}_k$. $\square$

Remark 5.2. If there exist more than $\aleph_0$ pairwise disjoint sets not in $\mathcal{A}$, then $\mathcal{A}$ is an inseparable algebra. However, it is not true that for every inseparable algebra there exist $\aleph_1$ pairwise disjoint sets not members of it. Indeed, let ν be a σ-additive extension of

Lebesgue measure defined on all subsets of the closed interval $[0,1]$ (see [So]). Define a two-valued measure μ on $[0,1]$ as follows: $\mu(M) = 0$ iff $\nu(M) = 0$. The algebra of all μ-measurable sets is inseparable, but there do not exist $\aleph_1$ pairwise disjoint μ-nonmeasurable sets.

After the next definition we shall be able to prepare immediately for the proof of Theorem 5.3.

Definition 5.5. A separable algebra is said to be strictly separable if it is not simple.

Let $\mathcal{A}$ be a strictly separable almost σ-algebra. There exists $Z \subset \beta X$ such that $|Z| = \aleph_0$ and $\overline{Z} \supset \ker \mathcal{A}$. Let

$$Z' = \{z \in Z \mid z \text{ is an irregular point}\},$$
$$Z'' = Z \backslash Z'.$$

It is obvious that all points Z'' are regular, $\overline{Z''} \supset \ker \mathcal{A}\backslash\overline{Z'} \neq \emptyset$, $Z'' \subset \beta X\backslash X$, and for every point $z \in Z''$ one can construct a function f, continuous on βX, such that $f(z) = 0$ and $f(x) > 0$ for $x \in X$. Let $U \subset X$, $Z' \cap \overline{U} = \emptyset$, $\overline{U} \cap \ker \mathcal{A} \neq \emptyset$,

$$Z_U = Z \cap \overline{U} = Z'' \cap \overline{U}.$$

Since $\mathcal{A}$ is an almost σ-algebra, consider a two-valued measure $\mu_{\mathcal{A}}^U$ defined on X by putting $\mu_{\mathcal{A}}^U(M) = 0$ iff $\mathcal{P}(M \cap U) \subset \mathcal{A}$. Obviously, $\mu_{\mathcal{A}}^U(M) = 0$ if and only if

$$\overline{M \cap U} \cap \ker \mathcal{A} = \emptyset.$$

Now consider a countable sequence of strictly separable almost σ-algebras $\mathcal{A}_1,\ldots,\mathcal{A}_n$, $\ldots$. For each algebra $\mathcal{A}_n$ consider the corresponding sets U_n, Z_{U_n} and the two-valued measure $\mu_{\mathcal{A}_n}^{U_n}$. Let $Y = \bigcup_n Z_{U_n}$ and

$$Y = \{y_1,\ldots,y_k,\ldots\}.$$

Obviously, $|Z_{U_n}| = \aleph_0$ and therefore $|Y| = \aleph_0$. For every point y_k there exists a continuous function f_k on βX such that $f_k(y_k) = 0$ and $f_k(x) > 0$ for $x \in X$. One can establish a one-to-one correspondence between $2^{\aleph_0}$ and the set of all countable sequences of positive real numbers. If $\iota \in 2^{\aleph_0}$ corresponds to a sequence $\alpha_1, \ldots, \alpha_k, \ldots$, we define

$$P_\iota = \{x \in X \mid f_1(x) \geq \alpha_1, \ldots, f_k(x) \geq \alpha_k, \ldots\}.$$

If ι_0 is the first element in $2^{\aleph_0}$, we put

$$Q_{\iota_0} = P_{\iota_0}.\ {}^{19}$$

If $\iota > \iota_0$, we put

$$Q_\iota = P_\iota \setminus \bigcup_{\lambda < \iota} P_\lambda.$$

For every n we define a two-valued measure μ_n on $2^{\aleph_0}$: $\mu_n(M) = 0$ iff

$$\mu_{\mathcal{A}_n}^{U_n}\left(\bigcup_{\iota \in M} Q_\iota\right) = 0.$$

By the Gitik-Shelah theorem, there exist pairwise disjoint sets

$$B_1, \ldots, B_n, \ldots \subset 2^{\aleph_0}$$

such that B_n is μ_n-nonmeasurable. Let

$$B_n^* = \bigcup_{\iota \in B_n} Q_\iota.$$

Clearly, $B_1^*, \ldots, B_n^*, \ldots$ are pairwise disjoint sets and

$$|\overline{B_n^* \cap U_n} \cap \ker \mathcal{A}_n| \geq \aleph_0.\ {}^{20}$$

Obviously, if C is any set of irregular points and $|C| \leq \aleph_0$, we can choose $U_1, \ldots, U_n, \ldots$ so that $C \cap \overline{U}_n = \emptyset$. It is also obvious that there exist $\aleph_0$ pairwise disjoint subsets of $B_n^* \cap U_n$ each of which is not a member of $\mathcal{A}_n$.

---

[19] By $2^{\aleph_0}$ we mean, as usual, a well-ordered set of the cardinality of the continuum.

[20] It is rather easy to prove that

$$|\overline{B_n^* \cap U_n} \cap \ker \mathcal{A}_n| = k = 2^{2^{\aleph_0}}.$$

Obviously, $k \leq 2^{2^{\aleph_0}}$. We now consider a discrete set $M \subset \overline{B_n^* \cap U_n} \cap \ker \mathcal{A}_n$, $|M| = \aleph_0$. We shall assume that there exist sets of ultrafilters M', M'' such that $M' \subset M$, $|M'| = \aleph_0$, for each $a \in M'$ there exists a $\mathcal{A}_n$-similar ultrafilter $b \in M''$, and $\overline{M'} \cap \overline{M''} = \emptyset$. Then $\overline{M'} \subset \ker \mathcal{A}_n$ (see footnote 12 in Section 3). But if no such sets M', M'' exist, there exists a set $M' \subset M$ such that $|M'| = \aleph_0$ and all ultrafilters in M' are $\mathcal{A}_n$-special. It is obvious that also in that case $\overline{M'} \subset \ker \mathcal{A}_n$. Thus, always

$$\overline{M'} \subset \ker \mathcal{A}_n,$$
$$|\overline{M'}| = |\beta N| = 2^{2^{\aleph_0}}.$$

Theorem 5.3. *Consider a countable sequence of almost σ-algebras $\mathcal{A}_1, \ldots, \mathcal{A}_k, \ldots$. Suppose there exists a set Q such that, if $\mathcal{A}_k$ is simple, then $Q \notin \mathcal{A}_k$. Then there exists a set that is not a member of any of these algebras.*

Proof. By Theorem 5.2, we may assume that $\mathcal{A}_1, \ldots, \mathcal{A}_k, \ldots$ are separable algebras. Let $\mathcal{A}_{k_1}, \ldots \mathcal{A}_{k_n}, \ldots$ be the sequence of all strictly separable algebras among $\mathcal{A}_1, \ldots, \mathcal{A}_k, \ldots$. Let C be a set of irregular points, $|C| \leq \aleph_0$, and $\overline{C} \supset \ker \mathcal{A}_k$ if $\mathcal{A}_k$ is a simple algebra. It follows from our previous arguments that we can construct a matrix of pairwise disjoint sets

$$\begin{pmatrix} U_1^1 & & \\ U_1^1 & U_2^2 & \\ \cdots\cdots\cdots\cdots & & \\ U_1^n & \cdots & U_n^n \\ \cdots\cdots\cdots\cdots & & \end{pmatrix}$$

such that $U_i^n \notin \mathcal{A}_{k_n}$ and

$$\overline{C} \cap \overline{\bigcup_{n,i} U_i^n} = \emptyset.$$

Reasoning just as in the proof of Theorem 5.2, we arrive at the proof of Theorem 5.3. $\square$

Remark 5.3. Previously, we considered the measures μ_n on $2^{\aleph_0}$. Clearly, the kernel of the algebra of all μ_n-measurable sets is an outer separable subset of $\beta 2^{\aleph_0}$. (By $\beta 2^{\aleph_0}$ we mean, of course, the Čech compactification of the set $2^{\aleph_0}$ with the discrete topology.) Therefore, a necessary and sufficient condition for the existence of a strictly separable almost σ-algebra is that on a set of the cardinality of the continuum there exists a two-valued measure μ such that the algebra of all μ-measurable sets is separable. If $2^{\aleph_0} = \aleph_1$, then there exists no strictly separable almost σ-algebra (see the Introduction for a corollary from the

---

(The fact that $|M'| = |\beta N|$ is obvious; that $|\beta N| = 2^{2^{\aleph_0}}$ is a known fact from general topology, see [P],[M].) Therefore $k = 2^{2^{\aleph_0}}$. However, the knowledge that $k > \aleph_0$ does not make it any easier to achieve our goal – the proof of Theorem 5.3. The arguments just presented imply the following assertions:

(1) *There exists no algebra $\mathcal{A}$ such that*

$$\aleph_0 \leq |\ker \mathcal{A}| < 2^{2^{\aleph_0}}.$$

(2) $\ker \mathcal{A}$ *is a countably compact subset of βX.* (Recall that a topological space Y is countably compact if any countable open cover of Y can be refined to a finite subcover of Y.)

possibility of constructing Ulam's matrix). In a model in which there exists no strictly separable almost σ-algebra, Theorem 5.2 is a generalization of the Gitik-Shelah theorem. In such a model, moreover, Theorem 5.3 is an obvious corollary of Theorem 5.2. However, there exists a model in which there exists a two-valued measure μ on a set of the cardinality of the continuum such that the algebra of all μ-measurable sets is separable (see [BD]).

6. Proof of Theorem II

Definition 6.1. The set

$$\{a \in \ker \mathcal{A} \mid a \text{ is an irregular point}\}$$

is called the spectrum of the algebra $\mathcal{A}$, denoted by $\mathrm{sp}\mathcal{A}$.

If $\mathcal{A}$ is a separable algebra, then $|\mathrm{sp}\mathcal{A}| \leq \aleph_0$.

Lemma 6.1. *If $\mathcal{A}$ is a simple almost σ-algebra, then $\overline{\mathrm{sp}\mathcal{A}} \supset \ker \mathcal{A}$.*

Proof. Suppose the contrary. Then there exist $U \subset X$ and a countable set of irregular points H such that

$$\overline{U} \supset \overline{H} \supset (\overline{U} \cap \ker \mathcal{A}) \neq \emptyset,$$

$$H \cap \ker \mathcal{A} = \emptyset.$$

Let

$$H = \{h_1, \ldots, h_n, \ldots\}, H_n = H \backslash \{h_n\}.$$

It is obvious that $\overline{H}_n \supset (\overline{U} \cap \ker \mathcal{A})$, $h_n \notin \overline{H}_n$. Hence there exists $U_n \in h_n$ such that $\mathcal{P}(U_n) \subset \mathcal{A}$. Thus all elements of $\mathcal{P}(\bigcup_n U_n)$ are members of $\mathcal{A}$. But this contradicts the fact that

$$\overline{\bigcup_n U_n} \supset \overline{H} \supset (\overline{U} \cap \ker \mathcal{A}) \neq \emptyset. \ \square$$

Definition 6.2. An algebra $\mathcal{A}$ is said to be ω-saturated if there do not exist infinitely many pairwise disjoint sets not member of $\mathcal{A}$.[21]

---

[21] The concept of a saturated algebra is not new (see, e.g., [G]).

An algebra $\mathcal{A}$ is ω-saturated if and only if $|\ker\mathcal{A}| < \aleph_0$. If $\mathcal{A} = \mathcal{P}(X)$, then $\mathcal{A}$ is ω-saturated and $\ker\mathcal{A} = \emptyset$. If $\mathcal{A}$ is ω-saturated, then the maximum number of pairwise disjoint sets not in $\mathcal{A}$ is $|\ker\mathcal{A}|$. An ω-saturated almost σ-algebra $\mathcal{A}$ is a σ-algebra and $\mathrm{sp}\mathcal{A} = \ker\mathcal{A}$.

Definition 6.3. A simple algebra $\mathcal{A}$ which is not ω-saturated will be called a strictly simple algebra.

The following assertion is obvious:

Assertion. *An almost σ-algebra $\mathcal{A}$ is strictly simple if and only if there exist $\aleph_0$ pairwise disjoint sets $M_1, \ldots, M_k, \ldots$ such that*

(a) $M_k \notin \mathcal{A}$;

(b) *if* $M', M'' \subset M_k$, $M' \cap M'' = \emptyset$, $M' \notin \mathcal{A}$, *then* $M'' \in \mathcal{A}$;

(c) *if* $M \subset X \backslash \bigcup_k M_k$, *then* $M \in \mathcal{A}$.

Definition 6.4. An algebra $\mathcal{A}$ majorizes an algebra $\mathcal{B}$ (notation: $\mathcal{A} > \mathcal{B}$) if:

(a) every $\mathcal{A}$-special ultrafilter is $\mathcal{B}$-special;

(b) every pair of $\mathcal{A}$-similar ultrafilters is a pair of $\mathcal{B}$-similar ultrafilters.

We can now proceed to

The proof of Theorem II. By Theorem 5.3, we may assume that all algebras $\mathcal{A}_k$ are simple. Let

$$Z = \bigcup_k \mathrm{sp}\mathcal{A}_k.$$

For each $\mathcal{A}_k$ there are two possible cases:

(1) there exists an ω-saturated σ-algebra $\mathcal{A}'_k$ such that $\mathcal{A}'_k > \mathcal{A}_k$, $\mathcal{A}'_k \neq \mathcal{P}(X)$, and if $k \neq 2$, then

$$|\ker\mathcal{A}'_k| > \frac{4}{3}(k-1);$$

(2) Case (1) does not occur.

In Case (1), $\ker\mathcal{A}'_k \subset Z$. In Case (2) we said $\mathcal{A}_k$ *is not reducible to an ω-saturated algebra.* For each $\mathcal{A}_\ell$ which is not reducible to an ω-saturated algebra, there is an infinite set $Z_\ell \subset Z$ such that if $z \in Z_\ell$, z has an $\mathcal{A}_\ell$-similar ultrafilter $c \in \overline{Z}_\ell \backslash Z_\ell$. It will suffice to contruct a set that is not a member of all the algebras $\mathcal{A}'_k$ and all $\mathcal{A}_\ell$ which are not reducible to ω-saturated algebras.

There are three possibilities:

I. There exist finitely many algebras $\mathcal{A}'_k$;

II. There exist infinitely many algebras $\mathcal{A}'_k$, and infinitely many algebras $\mathcal{A}_\ell$ which are not reducible to ω-saturated algebras;

III. There exist finitely many algebras $\mathcal{A}_\ell$ which are not reducible to ω-saturated algebras.

Case I. Suppose we have n algebras $\mathcal{A}'_k$. By the first part of Theorem I, consider the corresponding sets S_n, T_n (see Section 4). We know that $|S_n| \leq n$, $|T_n| \leq n$; $S_n, T_n \subset Z$. Let $\mathcal{A}_{\ell_1}, \ldots, \mathcal{A}_{\ell_m}, \ldots$ be all the algebras which are not reducible to ω-saturated algebras, and let

$$B = \begin{pmatrix} b_1^1 & b_2^1 & \\ \cdots\cdots\cdots\cdots \\ b_1^m & \cdots & b_{m+1}^m \\ \cdots\cdots\cdots\cdots \end{pmatrix}$$

be a matrix of pairwise distinct points of Z such that there exist $\mathcal{A}_{\ell_m}$-similar ultrafilters b_i^m, c_i^m and $c_i^m \in \overline{Z} \backslash Z$. Let $C = \{c_i^m\}$. In Case I we assume that $b_i^m \notin T_n$. Since all the b_i^m are irregular points, it follows from Lemma 5.1 that there exists a sequence

$$B^* = \{b_{p_1}^1, \ldots, b_{p_m}^m, \ldots\}$$

such that

$$\overline{B^*} \cap \overline{C} = \emptyset.$$

If $Q \subset X$, $S_n \cup B^* \subset \overline{Q}$, and $\overline{Q} \cap (T_n \cup C) = \emptyset$, then Q is not a member of all the algebras $\mathcal{A}_k$.

Case II. Let $\mathcal{A}'_{n_1}, \ldots, \mathcal{A}'_{n_k}, \ldots,$ $n_1 < n_2 < \ldots < n_k < \ldots$, be all the algebras $\mathcal{A}'_k$, and $\mathcal{A}_{\ell_1}, \ldots, \mathcal{A}_{\ell_m}, \ldots$ all the algebras which are not reducible to ω-saturated algebras. Obviously, $n_k - k \to \infty$. Consider a subsequence of natural numbers

$$\{n_{k_1} < n_{k_2} < \ldots < n_{k_m} < \ldots\} \subset \{n_1 < n_2 < \ldots < n_k < \ldots\}$$

such that

$$\frac{4}{3}(n_{k_m} - 1) - (m+1)(m+2) + 2 > \frac{4}{3}(k_m - 1).$$

As in Case I, consider the matrix B and set C. We impose an additional condition: $b_i^m \notin \ker \mathcal{A}'_{n_k}$ if $k < k_m$. If $k < k_1$ we put $\tilde{A}_k = \mathcal{A}'_{n_k}$. Let

$$k_m \leq k < k_{m+1},$$

and consider the algebra $\mathcal{A}'_{n_k}$ and the matrix

$$B^m = \begin{pmatrix} b_1^1 & b_2^1 \\ \cdots\cdots\cdots\cdots\cdots \\ b_1^m & \cdots & b_{m+1}^m \end{pmatrix}.$$

The matrix B^m contains $\frac{(m+1)(m+2)-2}{2}$ elements. We construct an algebra $\tilde{A}_k > \mathcal{A}'_{n_k}$, by defining $\ker \tilde{A}_k$:

(1) an ultrafilter a is $\tilde{A}_k$-special iff it is $\mathcal{A}'_{n_k}$-special and $a \notin B^m$;

(2) a, b are $\tilde{A}_k$-similar ultrafilters iff a, b are $\mathcal{A}'_{n_k}$-similar and $\{a, b\} \cap B^m = \emptyset$.

From the construction of $\ker \tilde{A}_k$ it follows immediately that an ultrafilter $a \in \ker \mathcal{A}'_{n_k} \setminus B^m$ is not in $\ker \tilde{A}_k$ if and only if a has an $\mathcal{A}'_{n_k}$-similar ultrafilter $b \in B^m$, and a does not have an $\mathcal{A}'_{n_k}$-similar ultrafilter $d \in \ker \mathcal{A}'_{n_k} \setminus B^m$. Therefore,

$$|\ker \mathcal{A}'_{n_k}| - |\ker \tilde{A}_k| \leq (m+1)(m+2) - 2.$$

But since

$$|\ker \mathcal{A}'_{n_k}| > \frac{4}{3}(n_k - 1)$$

and $k \geq k_m$, it follows that

$$|\ker \tilde{A}_k| > \frac{4}{3}(k - 1).$$

It is clear that

$$\ker \tilde{A}_k \subset Z, \quad B \cap \ker \tilde{A}_k = \emptyset.$$

By Remark 4.2, starting from the algebras $\tilde{A}_1, \ldots, \tilde{A}_k, \ldots$ we can construct corresponding sets S, T. Clearly, S, T, B are pairwise disjoint subsets of Z. As in Case I, we consider the sequence B^*. Let

$$B' = \{b^m_{p_m} \in B^* \mid c^m_{p_m} \in \overline{S}\},$$
$$S^* = S \cup (B^* \backslash B'),$$
$$T^* = Z \backslash S^*.$$

Since all the points of Z are irregular, there exists $Q \subset X$ such that $S^* \subset \overline{Q}, \overline{Q} \cap T^* = \emptyset$. It is easy to verify that Q is not a member of all the algebras $\mathcal{A}_k$.

Case III. Consider all the algebras $\mathcal{A}_{\ell_1}, \ldots, \mathcal{A}_{\ell_m}, \ell_1 < \ell_2 < \ldots < \ell_m$, which are not reducible to ω-saturated algebras. For each i, $1 \leq i \leq m$, choose an ultrafilter b_i such that

    (1) $b_i \in Z$;

    (2) $b_i \notin \ker \mathcal{A}'_k$ if $k < \ell_m$;[22]

    (3) $b_i \neq b_j$, $i \neq j$;

    (4) b_i has an $\mathcal{A}_{\ell_i}$-similar ultrafilter $c_i \in \overline{Z} \backslash Z$.

Let $\mathcal{A}'_{n_1}, \ldots, \mathcal{A}'_{n_k}, \ldots, n_1 < n_2 < \ldots < n_k < \ldots$, be all the algebras $\mathcal{A}'_k$. Considering the finite sequence of the algebras $\mathcal{A}'_{n_1}, \ldots, \mathcal{A}'_{n_k}$, we construct corresponding sets S_k, T_k by induction. If $n_k < \ell_m$, then it is obvious that $b_i \notin S_k \cup T_k$. If we had constructed a set not a member of all the algebras $\mathcal{A}'_{n_k}$, it would be sufficient if $|\ker \mathcal{A}'_{n_k}| > \frac{4}{3}(k-1)$, $k \neq 2$ (we know that $\mathcal{A}'_{n_2} \neq \mathcal{P}(X)$). But the important thing for us is that if $n_k > \ell_m$ and $n_k > 2$, then

$$|\ker \mathcal{A}'_{n_k}| > \frac{4}{3}(k + m - 1).$$

It is also important that if $\ell_m = 1$, then $b_1 \notin \ker \mathcal{A}'_2$ (see footnote 22). Therefore, we can require that

    (5) no bush of (S_k, T_k) contains b_i and b_j, $i \neq j$;

    (6) no bush of (S_k, T_k) contains b_i and an $\mathcal{A}'_{n_r}$-special ultrafilter s_r, $1 \leq r \leq k$,
        $S_k = \{s_1, \ldots, s_r, \ldots, s_k\}$;

---

[22] If there exist no algebras which are not reducible to ω-saturated algebras, Case III is greatly simplified and we shall not consider it. If $\ell_m = 1$, we demand that $b_1 \notin \ker \mathcal{A}'_2$.

(7) $b_i \notin T_k$.[23]

Suppose there is a natural number τ such that, for any k, any bush of (S_k, T_k) contains at most τ points. By Remark 4.2, we can construct corresponding sets $S, T \subset Z$ for the algebras $\mathcal{A}'_{n_1}, \ldots, \mathcal{A}'_{n_k}, \ldots$. Since $b_i \notin T_k$, for b_i there are two possibilities:

(a) $b_i \notin S \cup T$;

(b) $b_i \in S$; b_i is contained in a bush K_i of (S_{k_i}, T_{k_i}), and if $k > k_i$, then K_i is a bush of (S_k, T_k).

Let
$$B^- = \{b_i \in S \mid c_i \in \overline{S}\}.$$

If $b_i \in B^-$ we invert the bush K_i (this is possible thanks to property (6)). We now have new sets S, T. By properties (5) and (7), T contains b_i if and only if $b_i \in B^-$. Let
$$B^0 = \{b_i \mid c_i \in \overline{S}\},[24]$$
$$B^{00} = \{b_1, \ldots, b_m\} \backslash B^0,$$
$$S^\diamond = S \cup B^{00},$$
$$T^\diamond = Z \backslash S^\diamond.$$

---

[23] To get a better idea of why we can require conditions (5), (6), (7) to hold, let us imagine that we have constructed S_k and T_k satisfying these conditions, and proceed now to the construction of S_{k+1}, T_{k+1}. (It is obvious that such sets S_1 and T_1 can be constructed.) To that end, we consider algebras $\hat{\mathcal{A}}_1, \ldots, \hat{\mathcal{A}}_{m+k+1}$. If $1 \le i \le m$, then $\ker \hat{\mathcal{A}}_i = \{b_i\}$ (b_i is obviously an $\hat{\mathcal{A}}_i$-special ultrafilter). If $m < i \le m + k + 1$, then $\ker \hat{\mathcal{A}}_i = \ker \mathcal{A}'_{n_{i-m}}$. Clearly, there exists a set which is not a member of all the algebras $\hat{\mathcal{A}}_i$, $1 \le i \le m+k$. For these algebras we take the corresponding sets
$$\hat{S}_{m+k} = \{b_1, \ldots, b_m\} \cup S_k,$$
$$\hat{T}_{m+k} = T_k$$

(to avoid misunderstandings, we have labeled these sets with a circumflex ˆ). The algebras $\hat{\mathcal{A}}_1, \ldots, \hat{\mathcal{A}}_{m+k}$ and $\hat{\mathcal{A}}_{m+k+1}$ satisfy the conditions of Theorem 4.1; we now apply the arguments in the proof of that theorem to these algebras and construct sets $\hat{S}_{m+k+1}, \hat{T}_{m+k+1}$. Clearly, no bush of $(\hat{S}_{m+k+1}, \hat{T}_{m+k+1})$ contains b_i, b_j ($i \ne j$); no bush of $(\hat{S}_{m+k+1}, \hat{T}_{m+k+1})$ contains b_i and an $\hat{\mathcal{A}}_r$-special ultrafilter $\hat{s}_r$, $m < r \le m+k+1$, $\hat{S}_{m+k+1} = \{\hat{s}_1, \ldots, \hat{s}_r, \ldots, \hat{s}_{m+k+1}\}$; $b_i \notin \hat{T}_{m+k+1}$. It remains only to put
$$\hat{S}_{m+k+1} = \{b_1, \ldots, b_m, \hat{s}_{m+1}, \ldots, \hat{s}_{m+k+1}\},$$
$$S_{k+1} = \{\hat{s}_{m+1}, \ldots, \hat{s}_{m+k+1}\},$$
$$T_{k+1} = \hat{T}_{k+m+1}.$$

Clearly, S_{k+1}, T_{k+1} satisfy conditions (5),(6), (7).

[24] Obviously, $B^- \subset B^0$. In addition, B^0 may contain points not belonging to $S \cup T$.

If $Q \subset X$, $S^\diamond \subset \overline{Q}$, $\overline{Q} \cap T^\diamond = \emptyset$, then Q is not a member of all the algebras $\mathcal{A}_k$.

It remains to consider the possibility that there is no natural number τ such that, for any k, *any bush of* (S_k, T_k) *contains at most τ points.* Suppose we have already constructed sets S_{k^0}, T_{k^0}, and there exists a bush of S_{k^0}, T_{k^0}, containing not less than $6m + 4$ points. Then if $k > k^0$, there exist corresponding ultrafilters $q_1, \ldots, q_p \in \ker \mathcal{A}'_{n_k}$, $p > 2m$ (see Lemma 4.1).[25] Let $b_1^*, \ldots, b_m^* \subset Z$ be pairwise distinct ultrafilters, $b_i^* \in \ker \mathcal{A}_{\ell_i}$, $b_i^* \notin S_{k^0} \cup T_{k^0}$; and suppose b_i^* has an $\mathcal{A}_{\ell_i}$-similar ultrafilter $c_i^* \in \overline{Z} \backslash Z$. We are going to construct the sets S_{k^0+1}, T_{k^0+1}, considering the corresponding ultrafilters

$$q_1, \ldots, q_p \in \ker \mathcal{A}'_{n_{k^0+1}}$$

(knowledge of the proof of Theorem 4.1 is necessary to understand our arguments here). Among these ultrafilters there are $m + 1$, say $q_1, \ldots, q_{m+1}$, none of which is one of the b_i^*. Henceforth we shall consider only q_j's for which $1 \leq j \leq m + 1$. If there exists an $\mathcal{A}'_{n_{k^0+1}}$-special ultrafilter q_j, we use it to construct S_{k^0+1}, T_{k^0+1}. If none of the ultrafilters $q_1, \ldots, q_{m+1}$ is $\mathcal{A}'_{n_{k^0+1}}$-special, but there exist $\mathcal{A}'_{n_{k^0+1}}$-similar ultrafilters q_j, g such that

$$g \notin \{b_1^*, \ldots, b_m^*\},$$

we construct S_{k^0+1}, T_{k^0+1} using q_j, g. But if each of the ultrafilters q_j has an $\mathcal{A}'_{n_{k^0+1}}$-similar ultrafilter $b_{i_j}^*$, then there exist $\mathcal{A}'_{n_{k^0+1}}$-similar ultrafilters q_{j_1}, q_{j_2}. Using q_{j_1}, q_{j_2}, we construct S_{k^0+1}, T_{k^0+1}. Thus, on the assumption that a bush of S_{k^0}, T_{k^0} contains not less than $6m + 4$ points, we can assume that

$$(S_k \cup T_k) \cap \{b_1^*, \ldots, b_m^*\} = \emptyset$$

for any k. Under this assumption, by Remark 4.2, we can construct corresponding sets $S, T \subset Z$ for the algebras $\mathcal{A}'_{n_1}, \ldots, \mathcal{A}'_{n_k}, \ldots$. Obviously,

$$(S \cup T) \cap \{b_1^*, \ldots, b_m^*\} = \emptyset.$$

---

[25] See also Remark 4.4.

Let
$$B_0 = \{b_i^* \mid c_i^* \in \overline{S}\},$$
$$B_{00} = \{b_1^*, \ldots, b_m^*\} \backslash B_0,$$
$$S_\Diamond = S \cup B_{00},$$
$$T_\Diamond = Z \backslash S_\Diamond.$$

If $Q \subset X$, $S_\Diamond \subset \overline{Q}$, $\overline{Q} \cap T_\Diamond = \emptyset$, then Q is not a member of all the algebras $\mathcal{A}_k$. $\square$

The following example which, we believe, is of interest in itself shows that the almost σ-additivity condition in Theorem II is essential:

Example 6.1. There exists a countable sequence of algebras $\mathcal{A}_1, \ldots, \mathcal{A}_k, \ldots$ such that, for every k, there are more than $\frac{4}{3}(k-1)$ pairwise disjoint sets which are not members of $\mathcal{A}_k$. At the same time: (1) there is no set which is not a member of any $\mathcal{A}_k$; (2) for every k, there exists a set $Q_k \notin \mathcal{A}_m$ if $m \neq k$.

Construction. Let $|X| = \aleph_0$. Partition X into four pairwise disjoint countable sets:
$$A^1 = \{a_1^2, a_1^3, \ldots, a_1^n, \ldots\},$$
$$A^2 = \{a_2^2, a_2^3, \ldots, a_2^n, \ldots\},$$
$$B^1 = \{b_1^2, b_1^3, \ldots, b_1^n, \ldots\},$$
$$B^2 = \{b_2^2, b_2^3, \ldots, b_2^n, \ldots\}.$$

Obviously,
$$\beta X = \beta A^1 \cup \beta A^2 \cup \beta B^1 \cup \beta B^2,$$

and all these Čech compactifications are similar to one another and similar to βN. When we say that $\beta A^1, \beta A^2, \beta B^1, \beta B^2$ are similar, we are assuming that the points $a_1^n, a_2^n, b_1^n, b_2^n$ are similar ($n > 1$). Consider the following similar points:
$$a_1^1 \in \beta A^1 \backslash A^1, \ \ b_1^1 \in \beta A^2 \backslash A^2, \ \ a_2^1 \in \beta B^1 \backslash B^1, \ \ b_2^1 \in \beta B^2 \backslash B^2.$$

For every natural number n, consider the matrix
$$\begin{pmatrix} a_1^1 & a_2^1 & b_1^1 & b_2^1 \\ \cdots\cdots\cdots\cdots \\ a_1^n & a_2^n & b_1^n & b_1^n \end{pmatrix}$$

and, as in the proof of the second part of Theorem I (see Section 4), construct algebras $\mathcal{A}_1, \ldots, \mathcal{A}_{3n}$. We have thus constructed a countable sequence of algebras $\mathcal{A}_1, \ldots, \mathcal{A}_k, \ldots$. We know that for every k there are more than $\frac{4}{3}(k-1)$ pairwise disjoint sets which are not members of $\mathcal{A}_k$.

(1) *We claim that there exists no set Q which is not a member of any $\mathcal{A}_k$.* Indeed, suppose the contrary.[26] Then for any k, either

$$\{a_1^k, a_2^k\} \subset \overline{Q}$$

or

$$\{b_1^k, b_2^k\} \subset \overline{Q}.$$

In particular, either

$$\{a_1^1, a_2^1\} \subset \overline{Q}$$

or

$$\{b_1^1, b_2^1\} \subset \overline{Q}.$$

Let

$$\{a_1^1, a_2^1\} \subset \overline{Q}.$$

Then there exists n such that

$$\{a_1^n, b_1^n\} \subset \overline{Q},$$

and therefore $Q \in \mathcal{A}_{3n-2}$. Suppose now that

$$\{b_1^1, b_2^1\} \subset \overline{Q}.$$

Then there exists n such that

$$\{a_2^n, b_2^n\} \subset \overline{Q}.$$

At the same time, we know that either

$$\{a_1^n, a_2^n\} \subset \overline{Q},$$

---

[26] To understand the following reasoning, one should have a clear conception of the structure of the algebras $\mathcal{A}_1, \ldots, \mathcal{A}_k, \ldots$.

or

$$\{b_1^n, b_2^n\} \subset \overline{Q}.$$

If $\{a_1^n, a_2^n\} \subset \overline{Q}$, then $\{a_1^n, a_2^n, b_2^n\} \subset \overline{Q}$, and therefore $Q \in \mathcal{A}_{3n}$. But if $\{b_1^n, b_2^n\} \subset \overline{Q}$, then $\{a_2^n, b_1^n\} \subset \overline{Q}$, and therefore $Q \in \mathcal{A}_{3n-1}$.

(2) *We shall now prove that for every k there exists a set $Q_k \notin \mathcal{A}_m$ if $m \neq k$.* Letting n be some natural number, we shall construct the sets $Q_{3n-2}, Q_{3n-1}, Q_{3n}$. We shall require that for $k < n$, $a_1^k, a_2^k \in \overline{Q}_j$; $b_1^k, b_2^k \notin \overline{Q}_j$ ($j = 3n - 2, 3n - 1, 3n$). We shall also require that

$$\{a_1^n, a_2^n, b_1^n, b_2^n\} \cap \overline{Q}_{3n-2} = \{a_2^n\},$$

$$\{a_1^n, a_2^n, b_1^n, b_2^n\} \cap \overline{Q}_{3n-1} = \{a_1^n\},$$

$$\{a_1^n, a_2^n, b_1^n, b_2^n\} \cap \overline{Q}_{3n} = \{b_1^n\}.$$

It follows from the structure of the algebras $\mathcal{A}_1, \ldots, \mathcal{A}_k, \ldots$ that $Q_j \notin \mathcal{A}_m$ if $m \neq j$ ($j = 3n - 2, 3n - 1, 3n$). $\square$

Despite the fact that the proof of Theorem II is quite complicated, some important corollaries are easily proved. As an example, we prove the following generalization of Grzegorek's result cited in the Introduction (condition (ii)):

Corollary 6.1. *Let $|X| = 2^{\aleph_0}$, and let $\mathcal{A}_1, \ldots, \mathcal{A}_k, \ldots$ be a countable sequence of almost σ-algebras none of which is all of $\mathcal{P}(X)$, and assume that $\{x\} \in \mathcal{A}_k$ if $x \in X$. Then there exists a set Q which is not a member of any $\mathcal{A}_k$.*

Proof. Consider the sequence of two-valued measures $\mu, \ldots, \mu, \ldots$ defined on X by putting $\mu_k(M) = 0$ iff $\mathcal{P}(M) \subset \mathcal{A}_k$. By the Gitik-Shelah theorem, there exist pairwise disjoint sets $Q_1, \ldots, Q_k, \ldots$ such that Q_k is μ_k-nonmeasurable. Obviously,

$$|\overline{Q}_k \cap \ker \mathcal{A}_k| \geq \aleph_0$$

(see also footnote 20). Hence there exists a matrix of pairwise disjoint sets

$$\begin{pmatrix} U_1^1 & & \\ \cdots\cdots\cdots\cdots & & \\ U_1^k & \cdots & U_k^k \\ \cdots\cdots\cdots\cdots & & \end{pmatrix}$$

such that $U_i^k \notin \mathcal{A}_k$. The existence of a set Q not belonging to any $\mathcal{A}_k$ now follows from the simple Theorem 5.1. $\square$

Remark 6.1. If in the proof of Corollary 6.1, one takes Proposition 5.1 instead of Theorem 5.1, it is possible to choose the set Q such that $Q = \bigcup_k U_k$, where $U_1, \ldots, U_k, \ldots$ are pairwise disjoint sets, and $U_k \notin \mathcal{A}_k$.

We now go on to consider σ-algebras.

Lemma 6.2. *Let $\mathcal{A}$ be a simple σ-algebra, $a \in \mathrm{sp}\mathcal{A}$, a and a' $\mathcal{A}$-similar ultrafilters and $Q \in a'$. Then a has an $\mathcal{A}$-similar ultrafilter $b \in \mathrm{sp}\mathcal{A} \cap \overline{Q}$.*

Proof. The lemma is meaningful if $a' \in \overline{\mathrm{sp}\mathcal{A}} \backslash \mathrm{sp}\mathcal{A}$. Let $D \in \mathcal{A}, a, a'$;

$$ Z = (\mathrm{sp}\mathcal{A} \cap \overline{D \cap Q}) \backslash \{a\}. $$

It is clear that $Z \neq \emptyset$; let $Z = \{z_1, \ldots, z_n, \ldots\}$. Suppose a, z_n are not $\mathcal{A}$-similar ultrafilters for any n. Then for every n there exists $D_n \in z_n$ such that $D_n \notin a$, $D_n \in \mathcal{A}$. Clearly, $a \not\ni \bigcup_n D_n$ (a is an irregular point), $a' \ni \bigcup_n D_n$, and $\mathcal{A} \ni \bigcup_n D_n$ ($\mathcal{A}$ is a σ-algebra). But this contradicts the assumption that a, a' are $\mathcal{A}$-similar ultrafilters. $\square$

Remark 6.2. Suppose that all the algebras $\mathcal{A}_k$ in Theorem II are σ-algebras. By Theorem 5.3, we may assume that they are simple. Now, by Lemma 6.2, for any k there exists an ω-saturated σ-algebra $\mathcal{A}_k'$ such that $\mathcal{A}_k' > \mathcal{A}_k$, $\mathcal{A}_k' \neq \mathcal{P}(X)$, and if $k \neq 2$, then

$$ |\ker \mathcal{A}_k'| > \frac{4}{3}(k-1). $$

In this case, therefore, the proof of Theorem II is much simpler. Indeed, by Remark 4.2 we shall construct for algebras $\mathcal{A}_k'$ corresponding sets S and T such that all points of $S \cup T$ are irregular. Therefore $\overline{S} \cap \overline{T} = \emptyset$ and Theorem II is proved. The fact that there does not exist a set not belonging to all $\mathcal{A}_k'$ follows also from the general statement – Proposition 9.1.

To conclude this section, we present

Remark 6.3. Let $\mathcal{A}$ be a simple σ-algebra, $X \in \mathcal{A}$. By Lemma 6.2, $\mathrm{sp}\mathcal{A} = \bigcup_i V_i$, $|V_i| > 1$, $i = 1, 2, \ldots$. Any two distinct ultrafilters in V_i are $\mathcal{A}$-similar ultrafilters, and if $v \in V_i$,

$w \in V_j$ $(i \neq j)$, then v, w are not $\mathcal{A}$-similar ultrafilters. It follows from Lemma 6.2 that $\overline{V}_i$ is an $\mathcal{A}$-similar set (see Definition 3.3). We shall define a σ-additive measure μ such that $\mu(X) = 1$ and $\mathcal{A}$ is the algebra of all μ-measurable sets. A set M is measurable if and only if, for all i, either $V_i \subset \overline{M}$ or $V_i \cap \overline{M} = \emptyset$. If

$$|\{V_i\}| = n < \aleph_0$$

and M is a μ-measurable set,

$$\mu(M) = \frac{|\{i \mid V_i \subset \overline{M}\}|}{n}.$$

If $|\{V_i\}| = \aleph_0$ and M is a μ-measurable set, we denote

$$\chi_M(i) = \begin{cases} 0 & \text{if } V_i \cap \overline{M} = \emptyset, \\ 1 & \text{if } V_i \subset \overline{M}, \end{cases}$$

and then put

$$\mu(M) = \sum_i \frac{\chi_M(i)}{2^i}.$$

7. Improvement of Theorem II (Proof of Theorem II*)

Before the actual proof of Theorem II* we need five lemmas. The assertion of the first is obvious.

Lemma 7.1. *Let s be an irregular point, $U_1, \dots, U_k, \dots$ a countable sequence of sets each of which is in s. Then $s \ni \bigcap_k U_k$.*

Lemma 7.2. *Let s be an irregular point, $s \ni U$, and $\mathcal{A}_1, \dots, \mathcal{A}_k, \dots$ a countable sequence of algebras. Then there exists a set $V \in s$ such that $V \subset U$ and for every algebra $\mathcal{A}_k$ either $V \notin \mathcal{A}_k$ or $\mathcal{P}(V) \subset \mathcal{A}_k$.*

Proof. Suppose that there exists a set $U_k^* \in s \cap U$ for the algebra $\mathcal{A}_k$ such that either $\mathcal{P}(U_k^*) \subset \mathcal{A}_k$ or $W \notin \mathcal{A}_k$ for any $W \in s \cap U_k^*$. By Lemma 7.1, the intersection of all the sets U_k^* is a set $U^* \in s$. For each $\mathcal{A}_k$, one of the following conditions must hold:

(1) $W \notin \mathcal{A}_k$ for any $W \in s \cap U^*$;

(2) $\mathcal{P}(U^*) \subset \mathcal{A}_k$;

(3) for any set $U' \in s \cap U^*$, there exists a set $U'' \in s \cap U'$ such that

$$U'' \in \mathcal{A}_k, \ \overline{U''} \cap \ker \mathcal{A}_k \neq \emptyset.$$

We may assume without loss of generality that every algebra $\mathcal{A}_k$ satisfies condition (3). It is clear that we can construct a family of pairwise disjoint subsets of U^*,

$$U_1, V_1; \ldots; U_k, V_k; \ldots$$

such that

$$s \not\ni \bigcup_k (U_k \cup V_k),$$

and for every $\mathcal{A}_k$ there exist $\mathcal{A}_k$-similar ultrafilters s_k, t_k; $U_k \in s_k, V_k \in t_k$. Then the required set is

$$V = U^* \backslash \bigcup_k U_k. \ \square$$

Lemma 7.3. *Let $\mathcal{A}$ be an algebra; suppose that there exists a matrix of pairwise disjoint sets not members of $\mathcal{A}$:*

$$\begin{pmatrix} U_1^1 & & \\ U_1^2 & U_2^2 & \\ \cdots\cdots\cdots & \\ U_1^k & \cdots & U_k^k \\ \cdots\cdots\cdots \end{pmatrix}.$$

Then there exists a sequence of sets $U_{\beta_1}^1, U_{\beta_2}^2, \ldots, U_{\beta_k}^k, \ldots$ such that $\mathcal{A} \not\ni \bigcup_{k \in N^*} U_{\beta_k}^k$ for any nonempty subset of natural numbers N^*.

Proof. Let $U_{\beta_1}^1 = U_1^1$. We now proceed just as in the proof of Theorem 5.1. Assume that there exists a finite or infinite sequence

$$U_{\beta_1}^{\alpha_1^1}, U_{\beta_2}^{\alpha_2^1}, \ldots, U_{\beta_m}^{\alpha_m^1}, \ldots \ (1 < \alpha_1^1 < \alpha_2^1 < \ldots < \alpha_m^1 < \ldots)$$

such that

$$\mathcal{A} \ni U_{\beta_1}^1 \cup \bigcup_m U_{\beta_m^1}^{\alpha_m^1}.$$

In this case all the sets

$$U_{\beta_1}^{\alpha_1^1}, U_{\beta_2}^{\alpha_2^1}, \ldots, U_{\beta_m}^{\alpha_m^1}, \ldots$$

are called distinguished. From two sets U_1^2, U_2^2 choose one, say U_2^2, which is not distinguished. Put $U_{\beta_2}^2 = U_2^2$. Assume that there exists a finite or infinite sequence

$$U_{\beta_1^2}^{\alpha_1^2}, U_{\beta_2^2}^{\alpha_2^2}, \ldots, U_{\beta_m^2}^{\alpha_m^2}, \ldots \quad (2 < \alpha_1^2 < \alpha_2^2 < \ldots < \alpha_m^2 < \ldots)$$

such that

$$\mathcal{A} \ni U_{\beta_2}^2 \cup \bigcup_m U_{\beta_m^2}^{\alpha_m^2}.$$

In this case all the sets

$$U_{\beta_1^2}^{\alpha_1^2}, U_{\beta_2^2}^{\alpha_2^2}, \ldots, U_{\beta_m^2}^{\alpha_m^2}, \ldots$$

are called distinguished. Continuing this line of reasoning, we construct undistinguished sets

$$U_{\beta_3}^3, U_{\beta_4}^4, \ldots, U_{\beta_k}^k, \ldots \; .$$

Let N^* be a nonempty subset of natural numbers, k^* the least number in N^*. Clearly,

$$\mathcal{A} \not\ni U_{\beta_{k^*}}^{k^*} \cup \bigcup_{k \in N^* \setminus \{k^*\}} U_{\rho_k}^{\gamma_k}$$

if none of the sets $U_{\rho_k}^{\gamma_k}$ is distinguished and $\gamma_k \geq k^*$. In particular,

$$\mathcal{A} \not\ni \bigcup_{k \in N^*} U_{\beta_k}^k. \; \square$$

The next lemma is a refinement of Theorem 5.1.

Lemma 7.4. *Consider a countable sequence of algebras* $\mathcal{A}_1, \ldots, \mathcal{A}_k, \ldots$. *Assume that there is a matrix*

$$\begin{pmatrix} U_1^1 & & \\ U_1^2 & U_2^2 & \\ \cdots\cdots\cdots\cdots & & \\ U_1^k & \cdots & U_k^k \\ \cdots\cdots\cdots\cdots & & \end{pmatrix}$$

of pairwise disjoint sets such that $U_i^k \notin \mathcal{A}_k$. *Let* $W \subset X \setminus \bigcup_{k,i} U_i^k$. *Then there exists a family of sets* $\mathfrak{U}$ *which contains either the single set* W, *or* W *plus a finite or infinite* sequence of sets $U^{\alpha_1}_{\beta_1}, U^{\alpha_2}_{\beta_2}, \ldots, U^{\alpha_i}_{\beta_i}, \ldots, \alpha_1 < \alpha_2 < \ldots < \alpha_i < \ldots$. Moreover, this family can be constructed in such a way that for each $\mathcal{A}_k$ there exists a set $V_k \in \mathfrak{U}$ such that $\mathcal{A}_k \not\ni V_k \cup \cup \mathfrak{U}'$ for any subfamily $\mathfrak{U}' \subset \mathfrak{U}$; if $k = \alpha_i$, then $V_k = U^{\alpha_i}_{\beta_i}$.

Proof. For every U^k_i, we consider either an $\mathcal{A}_k$-special ultrafilter s^k_i or $\mathcal{A}_k$-similar ultrafilters s^k_i, t^k_i; $U^k_i \in s^k_i$; in the second case, $U^k_i \notin t^k_i$.

We begin with the set U^1_1. If there is no t^1_1, proceed to the sets U^2_1, U^2_2. Otherwise, there are three possibilities:

    (1) $W \in t^1_1$;

    (2) there exists a sequence $U^2_{\ell^1_2}, U^3_{\ell^1_3}, \ldots, U^k_{\ell^1_k}, \ldots$ such that $t^1_1 \ni \bigcup_{k>1} U^k_{\ell^1_k}$;

    (3) neither of the above.

In the first case we call U^1_1 a distinguished set and put $V_1 = W$; in the second, all the sets $U^k_{\ell^1_k}$ are called distinguished sets. In the second and third cases we put $V_1 = U^1_1$.

Consider the sets U^2_1, U^2_2. At least one of them, say U^2_2, is not distinguished. Call the set U^2_1 distinguished. If there is no t^2_2, proceed to the sets U^3_1, U^3_2, U^3_3. If there is an ultrafilter t^2_2, there are four possibilities:

    (1) $W \in t^2_2$;

    (2) there exists a sequence $U^3_{\ell^2_3}, U^4_{\ell^2_4}, \ldots, U^k_{\ell^2_k}, \ldots$ such that $t^2_2 \ni \bigcup_{k>2} U^k_{\ell^2_k}$;

    (3) $U^1_1 \in t^2_2$;

    (4) none of the above.

In the first case we call U^2_2 a distinguished set and put $V_2 = W$; in the second, all the sets $U^k_{\ell^2_k}$ are called distinguished sets. In the third case, if U^1_1 is not distinguished, we call U^2_2 distinguished and put $V_2 = U^1_1$. If U^2_2 is not a distinguished set, then $V_2 = U^2_2$.

Continuing the construction, we consider the sets

$$U^r_1, \ldots, U^r_r.$$

There exists at least one undistinguished set U^r_m. If $i \neq m$, we call U^r_i a distinguished set. If there is no ultrafilter t^r_m, proceed to the sets $U^{r+1}_1, \ldots, U^{r+1}_{r+1}$. Otherwise, there are four possibilities:

    (1) $W \in t^r_m$;

(2) there exists a sequence $U^{r+1}_{\ell^r_{r+1}}, U^{r+2}_{\ell^r_{r+2}}, \ldots, U^k_{\ell^r_k}, \ldots$ such that $t^r_m \ni \bigcup_{k>r} U^k_{\ell^r_k}$;

(3) $U^n_p \in t^r_m$, $n < r$;

(4) none of the above.

In the first case we call U^r_m a distinguished set and put $V_r = W$; in the second, all the sets $U^k_{\ell^r_k}$ are called distinguished sets. In the third case, if U^n_p is not distinguished, we call U^r_m distinguished and put $V_r = U^n_p$. If U^r_m is not a distinguished set, then $V_r = U^r_m$.

Consider all rows of the matrix of sets specified in the theorem. The family $\mathfrak{U}$ consists of the set W and all undistinguished sets U^k_i, if such exist. $\square$

The next lemma is a generalization and refinement of Theorem II in the case that the algebras $\mathcal{A}_k$ are simple:

Lemma 7.5. *Consider a countable sequence of almost σ-algebras $\mathcal{A}_1, \ldots, \mathcal{A}_k, \ldots$ such that for every k there exists a nonempty finite set of irregular points $H_k \subset \ker \mathcal{A}_k$ and $|H_k| > \frac{4}{3}(k-1)$ if $k \neq 2$. Then there exists $H \subset \bigcup_k H_k$ such that for any $\mathcal{A}_k$ at least one of the following two conditions holds:*

(1) H contains an $\mathcal{A}_k$-special ultrafilter;

(2) there exist $\mathcal{A}_k$-similar ultrafilters s_k, t_k such that $s_k \in \overline{H}$, $t_k \notin \overline{H}$.

Proof. Define $Z = \bigcup_k H_k$. If $\mathcal{A}_k$ satisfies the following two conditions, we assign it a corresponding ω-saturated σ-algebra $\mathcal{A}^0_k$. The conditions are:

(a) there exists a nonempty finite set $Z^*_k \subset Z$ such that

$$|Z^*_k| > \frac{4}{3}(k-1)$$

if $k \neq 2$;

(b) every $z \in Z^*_k$ satisfies one of the following three conditions:

(b$_1$) z is an $\mathcal{A}_k$-special ultrafilter;

(b$_2$) z has an $\mathcal{A}_k$-similar ultrafilter in Z^*_k;

(b_3) if z satisfies neither (b_1) nor (b_2), then it has an $\mathcal{A}_k$-similar ultrafilter
in $\beta X \backslash \overline{Z}$.

We define an ultrafilter z to be $\mathcal{A}_k^0$-special if $z \in Z_k^*$ and it satisfies either (b_1) or (b_3). Similarly, ultrafilters a, b ($a \neq b$) will be called $\mathcal{A}_k^0$-similar if $a, b \in Z_k^*$ and they are $\mathcal{A}_k$-similar.

If the algebra $\mathcal{A}_k$ is such that the algebra $\mathcal{A}_k^0$ cannot be constructed, we say that $\mathcal{A}_k$ is Z-irreducible. If $\mathcal{A}_k$ is a Z-irreducible algebra, there exists an infinite set $Z_k \subset Z \cap \ker \mathcal{A}_k$ such that no ultrafilter of Z_k is $\mathcal{A}_k$-special and all $\mathcal{A}_k$-similar ultrafilters of each ultrafilter of Z_k are elements of $\overline{Z}_k \backslash Z_k$.

Just as in the proof of Theorem II, we consider three cases:

I. There exist finitely many algebras $\mathcal{A}_k^0$;

II. There exist infinitely many algebras $\mathcal{A}_k^0$, and infinitely many Z-irreducible algebras $\mathcal{A}_\ell$;

III. There exists finitely many Z-irreducible algebras $\mathcal{A}_\ell$.

The rest of the arguments and the notations are almost exactly the same as in the proof of Theorem II. The only differences are as follows:

Instead of the algebra $\mathcal{A}_k'$, we consider $\mathcal{A}_k^0$.

Instead of algebras which are not reducible to ω-saturated algebras, we consider Z-irreducible algebras.

It remains only to define

in Case I:

$$H = S_n \cup B^*;$$

in Case II:

$$H = S^*;$$

in Case III: either

$$H = S^\diamond$$

or

$$H = S_\diamond. \quad \square$$

We can now proceed to the

Proof of Theorem II*.

Step I. To each set Q_i^k we associate (if possible) an ultrafilter

$$s_i^k \in \overline{Q_i^k} \cap \ker \mathcal{A}_k$$

such that

$$s_i^k \notin \overline{\ker \mathcal{A}_k \backslash \{s_i^k\}}.$$

Obviously, s_i^k is an irregular point.

The set Q_i^k is called *a set of the first kind* if, for all sets

$$Q_1^k, \ldots, Q_{m_k}^k$$

there exist ultrafilters

$$s_1^k, \ldots, s_{m_k}^k.$$

Q_i^k is called *a set of the second kind* if it is not a set of the first kind, but there exists an ultrafilter s_i^k.

Q_i^k is called *a set of the third kind* if it is neither of the first or of the second kind.

Obviously, if Q_i^k is a set of the third kind, then

$$|\overline{Q_i^k} \cap \ker \mathcal{A}_k| \geq 2^{2^{\aleph_0}}$$

(see footnote 20).

We introduce the following notation:

Z^* the set of all ultrafilters s_i^k;

Z the set of all ultrafilters s_i^k for sets of the first kind;

N^0 the set of all k such that the sets Q_i^k are sets of the first kind.

Let $H \subset Z$ be the set of ultrafilters corresponding to the family $\{\mathcal{A}_k\}_{k \in N^0}$ in accordance with Lemma 7.5 (here we set $H_k = \{s_1^k, \ldots, s_{m_k}^k\}$). For each $k \in N^0$, consider either an $\mathcal{A}_k$-special ultrafilter $s_k \in H$ or $\mathcal{A}_k$-similar ultrafilters s_k, t_k ($s_k \in \overline{H}$). Let T be the set of all t_k. We know that $\overline{H} \cap \overline{T} = \emptyset$.

For each $z \in Z^*$, consider a set $U_z \in z$. For each set Q_i^k of the third kind, consider a set $Q_{k,i} \subset Q_i^k; Q_{k,i} \notin \mathcal{A}_k$. By the same reasoning as in Section 5, we may assume that the sets $U_z, Q_{k,i}$ constitute a family of pairwise disjoint sets. In addition, we may demand that

(1) $T \cap \overline{\bigcup_{z \in H} U_z} = \emptyset$;

(2) if $Z^* \ni z = s_i^k$, then $U_z \subset Q_i^k$ (clearly, it may occur that $z = s_i^k = s_j^\ell$ and $k \notin \ell$);

(3) by Lemma 7.2, we may assume that for any set U_z and any algebra $\mathcal{A}_k$, either $U_z \notin \mathcal{A}_k$ or $\mathcal{P}(U_z) \subset \mathcal{A}_k$.[27]

We can already define some of the required sets. If $s_i^k = z \in Z^*$, then

$$\hat{Q}_i^k = U_z.$$

Step II. Let $t_k \in T$ and $t_k \ni Q_{m,i}$. We may assume that this is possible only when

$$t_k \in \overline{Q_{m,i} \cap \ker \mathcal{A}_m}$$

and $L_{m,i} = \overline{Q}_{m,i} \cap \ker \mathcal{A}_m$ is an outer separable set, all points of which are regular. We claim that in this case there exists a set $M_{m,i}^k \in t_k$ such that $M_{m,i}^k \subset Q_{m,i}$ and

$$\overline{M_{m,i}^k} \cap L_{m,i} \subset \overline{\ker \mathcal{A}_k}.$$

Indeed, consider a countable set $\Upsilon \subset \beta X$ such that

$$\overline{\Upsilon} \supset L_{m,i}.$$

Let

$$\Upsilon' = \{v \in \Upsilon \mid \text{there exists a set } O_v \in v \text{ such that } \mathcal{P}(O_v) \subset \mathcal{A}_k\}.$$

Since $\mathcal{A}_k$ is an almost σ-algebra, we have

$$\mathcal{A}_k \ni Q_{m,i} \cap \bigcup_{v \in \Upsilon'} O_v.$$

Let $t_k \in \overline{\Upsilon'}$. Since $s_k \in \overline{H}$, $H \cap \overline{Q}_{m,i} = \emptyset$ and

$$t_k \ni Q_{m,i} \cap \bigcup_{v \in \Upsilon'} O_v,$$

it follows that

$$\mathcal{A}_k \not\ni Q_{m,i} \cap \bigcup_{v \in \Upsilon'} O_v.$$

Therefore $t_k \notin \overline{\Upsilon'}$. It remains only to take a set $M_{m,i}^k \in t_k$ such that $M_{m,i}^k \subset Q_{m,i}$ and

$$\overline{\Upsilon'} \cap \overline{M_{m,i}^k} = \emptyset.$$

Now consider the two-valued measure $\mu_{\mathcal{A}_m}^{M_{m,i}^k}$. Clearly, there exist $\mu_{\mathcal{A}_m}^{M_{m,i}^k}$-nonmeasurable sets, and each such set contains a set which is simultaneously $\mathcal{A}_m$-nonmeasurable and $\mathcal{A}_k$-nonmeasurable.

Let

$$T_{m,i} = \{t_k \in T \mid t_k \ni Q_{m,i}\}.$$

Reasoning as in Section 5, we construct a family of pairwise disjoint subsets of $Q_{m,i}$,

$$Q'_{m,i}, \quad \{Q_{m,i}^k\}_{t_k \in T_{m,i}},$$

such that $Q'_{m,i}$ is $\mathcal{A}_m$-nonmeasurable, and $Q_{m,i}^k$ is both $\mathcal{A}_m$-nonmeasurable and $\mathcal{A}_k$-nonmeasurable.

Step III. For every algebra $\mathcal{A}_m$, $m \notin N^0$, let us consider one set Q_{m,i_m} and a corresponding set T_{m,i_m}. We can now define yet another part of the required sets. If

$$Q_{m,j} \neq Q_{m,i_m},$$

then

$$\hat{Q}_j^m = Q_{m,j}.$$

Let T^0 be the union of all sets T_{m,i_m}. By Lemma 5.1, we may assume that if $t \in T$ and

$$t \ni \bigcup_m Q_{m,i_m},$$

then t contains some Q_{m,i_m}, and therefore $t \in T^0$.

Let $\mathfrak{A}$ be the family of all algebras $\mathcal{A}_m$, $m \notin N^0$, and all algebras $\mathcal{A}_k$, $t_k \in T^0$. Consider the matrix of pairwise disjoint sets

$$\begin{pmatrix} U_1^1 & & \\ U_1^2 & U_2^2 & \\ \cdots\cdots\cdots\cdots & \\ U_1^p & \cdots & U_p^p \\ \cdots\cdots\cdots\cdots & \end{pmatrix}.$$

There is a one-to-one correspondence between the rows of this matrix and the algebras in $\mathfrak{A}$. Suppose that the p-th row corresponds to an algebra $\mathcal{A}$. Then $U_j^p \notin \mathcal{A}$. If $\mathcal{A} = \mathcal{A}_m$, $m \notin N^0$, then $U_j^p \subset Q_{m,i_m}$. If $\mathcal{A} = \mathcal{A}_k$, $t_k \in T_{m,i_m}$, then $Q_j^p \subset Q_{m,i_m}$. Let $\mathfrak{U}^{(m)}$ be the family of all sets U_j^p which are subsets of Q_{m,i_m}. By Lemma 7.3, using the sets $Q'_{m,i}$, $\{Q_{m,i}^k\}_{t_k \in T_{m,i}}$ defined in Step II, we may assume that for any nonempty subfamily $\mathfrak{U}' \subset \mathfrak{U}^{(m)}$,

$$\mathcal{A}_m \not\ni \cup\mathfrak{U}'.$$

Now consider the set $W = \bigcup_{z \in H} U_z$, the sets U_j^p and the family of algebras $\mathfrak{A}$. By Lemma 7.4, there exists a family of sets $\mathfrak{U}$ which contains either the single set W or the set W and a finite or infinite sequence of sets

$$U_{\beta_1}^{\alpha_1}, U_{\beta_2}^{\alpha_2}, \ldots, U_{\beta_i}^{\alpha_i}, \ldots, \alpha_1 < \alpha_2 < \ldots < \alpha_i < \ldots .$$

Moreover, for each algebra $\mathcal{A} \in \mathfrak{A}$ there is a set $V_{\mathcal{A}} \in \mathfrak{U}$ such that

$$\mathcal{A} \not\ni V_{\mathcal{A}} \cup \cup\mathfrak{U}^*$$

for an arbitrary subfamily $\mathfrak{U}^* \subset \mathfrak{U}$.

We can now complete our constructions. Denote the union of all sets in $\mathfrak{U}$ which are subsets of Q_{m,i_m} by $\hat{Q}_{i_m}^m$. If no such sets exist, then by definition

$$\hat{Q}_{i_m}^m = Q_{m,i_m}.$$

The required set Q will be $\cup\mathfrak{U}$. Clearly, $\hat{Q}_i^k \subset Q_i^k$ and $\hat{Q}_i^k \notin \mathcal{A}_k$; either $\hat{Q}_i^k = \hat{Q}_j^\ell$ or $\hat{Q}_i^k \cap \hat{Q}_j^\ell = \emptyset$. (If $\hat{Q}_i^k = \hat{Q}_j^\ell$ and $k \neq \ell$, then $\hat{Q}_i^k = \hat{Q}_j^\ell = U_z$.) It is clear that $Q \notin \mathcal{A}_k$ for all k. It remains only to prove that for each algebra $\mathcal{A}_k$ there exists a corresponding set $\hat{Q}_{\psi_k}^{\xi_k}$.

Indeed, if $\mathcal{A}_k \in \mathfrak{A}$ and $V_{\mathcal{A}_k} \subset Q_{m,i_m}$, then $\hat{Q}_{\psi_k}^{\xi_k} = \hat{Q}_{i_m}^{m}$. Consider the case in which $\mathcal{A}_k \in \mathfrak{A}$ and $V_{\mathcal{A}_k} = W$. Since $\mathcal{A}_k$ is an almost σ-algebra and for every $U_z \subset W$, either $U_z \notin \mathcal{A}_k$ or $\mathcal{P}(U_z) \subset \mathcal{A}_k$, there exists $U_{z_k} \notin \mathcal{A}_k$ $(z_k \in H)$. Now put $\hat{Q}_{\psi_k}^{\xi_k} = U_{z_k}$. Finally, consider the case $\mathcal{A}_k \notin \mathfrak{A}$. Here $W \notin \mathcal{A}_k$ and we again find $\hat{Q}_{\psi_k}^{\xi_k} = U_{z_k} \notin \mathcal{A}_k$ $(z_k \in H)$. □

Proceeding to the concluding part of this section, we observe that the proofs of the first part of Theorem I and Theorem 4.2 imply assertions similar to those of the first part of Theorem V and Theorem 10.2 (Theorem 10.2 is a refinement of the first part of Theorem V). The following result is an obvious corollary of the proof of Theorem 4.2:

Theorem. *Consider a finite sequence of algebras* $\mathcal{A}_1, \ldots, \mathcal{A}_n$. *Suppose that for each* k, $1 \leq k \leq n$, *we are given a finite sequence of pairwise disjoint sets* $Q_1^k, \ldots, Q_{m_k}^k \notin \mathcal{A}_k$, *where* $m_k > \frac{4}{3}(k-1)$ *if* $k \neq 2$. *Then for every set* Q_i^k *we can construct a set* $\hat{Q}_i^k \subset Q_i^k$ *such that* $\hat{Q}_i^k \notin \mathcal{A}_k$; *either* $\hat{Q}_i^k = \hat{Q}_j^\ell$ *or* $\hat{Q}_i^k \cap \hat{Q}_j^\ell = \emptyset$; *for every* k, $1 \leq k \leq n$, *there exists a set* $\hat{Q}_{\psi_k}^{\xi_k}$ *such that if* $\hat{Q}_{\psi_k}^{\xi_k} \subset U \subset \bigcup_{i=1}^{n} \hat{Q}_{\psi_i}^{\xi_i}$, *then* $U \notin \mathcal{A}_k$.[28]

The essential difference between this theorem and Theorem 10.2 is that here equality $\hat{Q}_{\psi_i}^{\xi_i} = \hat{Q}_{\psi_j}^{\xi_j}$ may occur even though $i \neq j$.

We now outline the arguments by virtue of which the proof of Theorem II*, applied to σ-algebras rather than almost σ-algebras, will yield a result similar to Theorem 12.5 (which is a refinement of Theorem VI). This result is of interest specifically because we are forced to consider σ-algebras and not almost σ-algebras.

If σ-algebras are considered, rather than almost σ-algebras, then, first, the construction of H is much simpler. Indeed, our original tool for that purpose was Lemma 7.5. But if the $\mathcal{A}_k$'s in the statement of the latter are σ-algebras, it follows from Lemma 6.2 that there is no need to consider Z-irreducible algebras. Therefore, we may assume that $s_k \in H$ (in the proof of Theorem II*, $s_k \in \overline{H}$). Second, suppose again that the p-th row $U_1^p, \ldots, U_p^p$ corresponds to an algebra $\mathcal{A} \in \mathfrak{A}$. If x, y are $\mathcal{A}$-similar ultrafilters, $U_j^p \in x$ and $y \in \overline{H}$, then there exist $\mathcal{A}$-similar ultrafilters x', y'; $U_j^p \in x'$; $y' \in H$.[29] Therefore, we may assume that for any set U_j^p there exists either an $\mathcal{A}$-special ultrafilter x_j^p or $\mathcal{A}$-similar ultrafilters x_j^p, y_j^p;

[28] In connection with the similarity between this theorem and the first part of Theorem V and Theorem 10.2, see Remark 2.2.

[29] See the proofs of Lemmas 6.2 and 8.1.

$U_j^p \in x_j^p$, $U_j^p \notin y_j^p$, and if $W \in y_j^p$, then $y_j^p = z_p \in H$.[30] Again consider the sets W, U_j^p and the family of algebras $\mathfrak{A}$. By Lemma 7.4, there exists a corresponding family of $\mathfrak{U}$ (here we use ultrafilters x_j^p, y_j^p; see the proof of Lemma 7.4). Let

$$\mathfrak{U}^* = \{U_z \mid z \in H\},$$

$$\mathfrak{U}^{**} = \mathfrak{U} \backslash \{W\}.$$

It is important that if $V_{\mathcal{A}} = W$, then there exists $y_j^p = z_p \in H$, and

$$\mathcal{A} \not\ni U_{z_p} \cup \cup \mathcal{V}$$

for an arbitrary subfamily $\mathcal{V} \subset \mathfrak{U}^* \cup \mathfrak{U}^{**}$.

This reasoned analysis of the proof of Theorem II* yields the following

Theorem 7.1. *Consider a countable sequence of σ-algebras $\mathcal{A}_1, \ldots, \mathcal{A}_k, \ldots$. Suppose that for each $\mathcal{A}_k$ we are given a finite sequence of pairwise disjoint sets $Q_1^k, \ldots, Q_{m_k}^k \notin \mathcal{A}_k$, where $m_k > \frac{4}{3}(k-1)$ if $k \neq 2$. Then for every set Q_i^k we can construct a set $\hat{Q}_i^k \subset Q_i^k$ such that $\hat{Q}_i^k \notin \mathcal{A}_k$; either $\hat{Q}_i^k = \hat{Q}_j^\ell$ or $\hat{Q}_i^k \cap \hat{Q}_j^\ell = \emptyset$; moreover, for each $\mathcal{A}_k$ there exists a set $\hat{Q}_{\psi_k}^{\xi_k}$ such that if $\hat{Q}_{\psi_k}^{\xi_k} \subset U \subset \bigcup_i \hat{Q}_{\psi_i}^{\xi_i}$, then $U \notin \mathcal{A}_k$.*

It is clear that Theorem 7.1 would be valid if we assumed that all the $\mathcal{A}_k$'s are almost σ-algebras, provided the sets U_z, $z \in H$, had not appeared in the proof. This makes it possible to prove the next theorem, whose proof does not involve similar sets. But first we need a definition:

Definition 7.1. Let $\mathcal{A}$ be an algebra. A set $M \subset X$ is said to be $\mathcal{A}$-saturated if there exist $\aleph_0$ pairwise disjoint subsets of M, none of which is a member of $\mathcal{A}$.

Theorem 7.2. *Let $\mathcal{A}_1, \ldots, \mathcal{A}_k, \ldots$ be a countable sequence of almost σ-algebras, and for each k let Q_k be a given $\mathcal{A}_k$-saturated set. Then there exists a sequence of pairwise disjoint sets $\hat{Q}_1, \ldots, \hat{Q}_k, \ldots$ such that $\hat{Q}_k \subset Q_k$ and $\hat{Q}_k \notin \mathcal{A}_k$; for each algebra $\mathcal{A}_k$ there exists a set $\hat{Q}_{\psi_k}$ such that if $\hat{Q}_{\psi_k} \subset U \subset \bigcup_i \hat{Q}_{\psi_i}$, then $U \notin \mathcal{A}_k$.*

---

[30]On the strength of the arguments that will be presented in Section 11, we may construct sets W, U_j^p and ultrafilters x_j^p, y_j^p such that $W \notin y_j^p$ (see Definition 11.3 and Lemma 11.4).

Proof. By the arguments set out in Section 5, we can construct a matrix of pairwise disjoint sets

$$\begin{pmatrix} U_1^1 & & \\ U_1^2 & U_2^2 & \\ \cdots\cdots\cdots\cdots & \\ U_1^k & \cdots & U_k^k \\ \cdots\cdots\cdots\cdots & \end{pmatrix}$$

such that $U_i^k \subset Q_k$ and $U_i^k \notin \mathcal{A}_k$. Now we apply Lemma 7.4 with $W = \emptyset$. Consider a corresponding sequence of sets $U_{\beta_1}^{\alpha_1}, U_{\beta_2}^{\alpha_2}, \ldots, U_{\beta_i}^{\alpha_i}, \ldots,$ $\alpha_1 < \alpha_2 < \ldots < \alpha_i < \ldots$ (see Lemma 7.4). If $k \neq \alpha_i$, then $\hat{Q}_k = U_1^k$; if $k = \alpha_i$, then $\hat{Q}_k = U_{\beta_i}^{\alpha_i}$. If $V_k = U_{\beta_i}^{\alpha_i}$ (see Lemma 7.4), then $\hat{Q}_{\psi_k} = U_{\beta_i}^{\alpha_i}$. $\square$

The next theorem has a direct bearing on the questions involved in refining Theorem VI, although it concerns almost σ-algebras. We first formulate a definition and provide a few words of explanation.

Definition 7.2. Let $\mathcal{A}$ be an algebra. A set $M \subset X$ is said to be strictly $\mathcal{A}$-saturated if, for any set $M' \subset M$ and an arbitrary countable sequence of ultrafilters $a_1, \ldots, a_k, \ldots$, there exists a set $M'' \subset M$ such that $M'' \notin \mathcal{A}$, $a_1, \ldots, a_k, \ldots$, and $M'' \subset M'$ if $M' \notin \mathcal{A}$.

Clearly, a set M is strictly $\mathcal{A}$-saturated if and only if $\overline{M} \cap \ker \mathcal{A} \neq \emptyset$, and $\overline{M'} \cap \ker \mathcal{A}$ is an outer separable set for any $M' \subset M$ if $\overline{M'} \cap \ker \mathcal{A} \neq \emptyset$. Suppose we are in a model in which there exists no strictly separable almost σ-algebra (e.g., if $2^{\aleph_0} = \aleph_1$), and $\mathcal{A}$ is an almost σ-algebra. Then a set $K \subset X$ is strictly $\mathcal{A}$-saturated if and only if K is $\mathcal{A}$-saturated, and any set $K' \subset K$, $K' \notin \mathcal{A}$, is $\mathcal{A}$-saturated.

Theorem 7.3. *Let $\mathcal{A}_1, \ldots, \mathcal{A}_k, \ldots$ be a countable sequence of almost σ-algebras, and for each k let Q_k be a given strictly $\mathcal{A}_k$-saturated set. Then there exists a sequence of pairwise disjoint sets $\hat{Q}_1, \ldots, \hat{Q}_k, \ldots$ such that $\hat{Q}_k \subset Q_k$ and if $\hat{Q}_k \subset U \subset \bigcup_i \hat{Q}_i$, then $U \notin \mathcal{A}_k$.*

Proof. As in the proof of Theorem 7.2 we consider a matrix of pairwise disjoint sets

$$\begin{pmatrix} U_1^1 & & \\ U_1^2 & U_2^2 & \\ \cdots\cdots\cdots\cdots & \\ U_1^k & \cdots & U_k^k \\ \cdots\cdots\cdots\cdots & \end{pmatrix}$$

such that $U_i^k \subset Q_k$ and $U_i^k \notin \mathcal{A}_k$. Clearly, the set

$$\overline{U_i^k} \cap \ker \mathcal{A}_k$$

is outer nonseparable. For each pair of sets U_i^k, U_p^m, $m > k$, let us consider an ultrafilter $\tau_{m,p}^{k,i} \ni U_i^k$ (if it exists) such that, for any set $V \in \tau_{m,p}^{k,i}$, there exists an outer nonseparable subset of $\overline{U_p^m} \cap \ker \mathcal{A}_m$ each of whose ultrafilters has an $\mathcal{A}_m$-similar ultrafilter that contains V. For each set U_i^k we consider the at most countable set of ultrafilters

$$T_i^k = \bigcup_{m>k} \{\tau_{m,p}^{k,i}\}.$$

As in the proof of Lemma 7.4, we consider for each U_i^k either an $\mathcal{A}_k$-special ultrafilter s_i^k or $\mathcal{A}_k$-similar ultrafilters s_i^k, t_i^k; $s_i^k \ni U_i^k$. However, unlike the parallel step in the proof of Lemma 7.4, these ultrafilters are chosen in a different way, and it may occur that $t_i^k \ni U_i^k$.

Let $s_1^1 \notin \overline{T_1^1}$. If s_1^1 is not an $\mathcal{A}_1$-special ultrafilter, we associate with it an $\mathcal{A}_1$-similar ultrafilter t_1^1. If $t_1^1 \ni U_1^1$, define

$$T_{1,1} = T_1^1 \cup \{t_1^1\}.$$

Otherwise, let $T_{1,1} = T_1^1$. Take a set $U_{1,1} \in s_1^1 \cap U_1^1$ such that

$$T_{1,1} \cap \overline{U}_{1,1} = \emptyset.$$

Take an $\mathcal{A}_2$-special ultrafilter $s_1^2 \notin \overline{T_1^2}$. If no such ultrafilter exists, take $\mathcal{A}_2$-similar ultrafilters s_1^2, t_1^2 such that

$$s_1^2 \notin \overline{T_1^2}, \ t_1^2 \not\ni U_1^1.$$

If such ultrafilters do not exist either, then the ultrafilter $\tau_{2,1}^{1,1}$ exists, and therefore there exist $\mathcal{A}_2$-similar ultrafilters s_1^2, t_1^2 such that

$$s_1^2 \notin \overline{T_1^2}, \ t_1^2 \ni U_1^1, \ t_1^2 \not\ni U_{1,1}.$$

If $t_1^2 \ni U_1^2$, define

$$T_{2,1} = T_1^2 \cup \{t_1^2\}.$$

Otherwise, $T_{2,1} = T_1^2$. Take a set $U_{2,1} \in s_1^2 \cap U_1^2$ such that

$$T_{2,1} \cap \overline{U}_{2,1} = \emptyset.$$

Proceeding in an analogous fashion, we choose the ultrafilter s_2^2, the ultrafilter t_2^2 (if necessary) and the set $U_{2,2}$.

Reasoning in this way for each set U_i^k, we shall have constructed

(a) a set $U_{k,i} \subset U_i^k$;

(b) either an $\mathcal{A}_k$-special ultrafilter s_i^k or $\mathcal{A}_k$-similar ultrafilter s_i^k, t_i^k; $s_i^k \ni U_{k,i}, t_i^k \not\ni U_{k,i}$;

it is important that $t_i^k \not\ni U_{\ell,j}$ if $\ell < k$.

The rest of the argument is exactly the same as in the proof of Lemma 7.4 with $W = \emptyset$ and the matrix

$$\begin{pmatrix} U_{1,1} & & \\ U_{2,1} & U_{2,2} & \\ \cdots\cdots\cdots\cdots\cdots & \\ U_{k,1} & \cdots & U_{k,k} \\ \cdots\cdots\cdots\cdots\cdots & \end{pmatrix}.$$

Since $t_i^k \not\ni U_{\ell,j}$ if $\ell < k$, it follows that each row of this matrix contains an undistinguished set U_{k,β_k} (see the proof of Lemma 7.4). It remains only to put $\hat{Q}_k = U_{k,\beta_k}$. $\square$

Remark 7.1. It will be an obvious conclusion from our reasoning in Sections 11, 12 that if the algebras in Theorem 7.2 are assumed to be not almost σ-algebras but σ-algebras, then the existence of sets $\hat{Q}_k$ can be established as in Theorem 7.3.

8. Proof of Theorems III[31] and IV

Lemma 8.1. *Let $\mathcal{A}$ be a σ-algebra; $Q_1, \ldots, Q_n, \ldots$ a countable sequence of pairwise disjoint sets; a, b are $\mathcal{A}$-similar ultrafilters; $Q_1 \in a$; $b \ni \bigcup_n Q_n$ and $Q_n \notin b$ for any n. Then there exist $\mathcal{A}$-similar ultrafilters a', b' such that $Q_1 \in a'$, $Q_n \in b'$ $(n > 1)$.*

Proof. Let $D \in \mathcal{A}$, a, b; $Q_n' = Q_n \cap D$. Suppose that if $Q_1' \in a'$, $Q_n' \in b'$, $n > 1$, then a', b' are not $\mathcal{A}$-similar ultrafilters. Then for every $n > 1$ there exists $U_n \in \mathcal{A}$ such that

---

[31] In this section, only the first part of Theorem III will be proved. We have already pointed out (in Section 2) that the second part of Theorem II was actually proved in the Introduction.

$Q'_n \subset U_n$, $Q'_1 \cap U_n = \emptyset$. Since $\mathcal{A}$ is a σ-algebra, $\mathcal{A} \ni U = \bigcup_{n>1} U_n$. On the other hand, $U \notin a$, $U \in b$ and therefore $U \notin \mathcal{A}$, a contradiction. $\square$

Remark 8.1. The proof of Lemma 8.1 is rather like the proof of Lemma 6.2.

Proof of the first part of Theorem III. With each of the sets U_i^k we associate either an $\mathcal{A}_k$-special ultrafilter $s_i^k \ni U_i^k$ or $\mathcal{A}_k$-similar ultrafilters s_i^k, t_i^k; $U_i^k \in s_i^k$, $U_i^k \notin t_i^k$. By Lemma 8.1, we may assume that either $U_p^m \in t_i^k$ or $t_i^k \not\supseteq \bigcup_{m,p} U_p^m$. If $k > 2$, we shall refer to the sets U_1^k, U_2^k as adjacent. We are now going to label some of the sets U_i^k with digits 0 or 1. If s_1^k is $\mathcal{A}_k$-special ($k \leq 2$), U_1^k is labeled 1. If at least one of the ultrafilters s_1^k, s_2^k is $\mathcal{A}_k$-special ($k > 2$), choose one set $U_{i_0}^k$ such that $s_{i_0}^k$ is $\mathcal{A}_k$-special; label this set 1. We shall assume that $U_{i_0}^k$ has no adjacent set, i.e., the adjacent set is not of the form U_i^k. If there exists a labeled set $U_{i(k)}^k$ for each k, let U denote the union of all labeled sets; obviously, $U \notin \mathcal{A}_k$ for all k. Otherwise, we shall consider finite or infinite sequences of sets

$$V_1 = U_{i_1}^{k_1}, \ V_2 = U_{i_2}^{k_2}, \ldots, V_n = U_{i_n}^{k_n}, \ldots$$

possessing the following properties:

(1) $U_{i_2}^{k_2} \in t_{i_1}^{k_1}$, $U_{i_3}^{k_3} \in t_{i_2}^{k_2}, \ldots, U_{i_{n+1}}^{k_{n+1}} \in t_{i_n}^{k_n}, \ldots$;

(2) all the sets V_n are distinct and none of them are adjacent;

(3) none of the sets V_n are labeled;

(4) V_n has no labeled adjacent set.

We call the sequence $V_1, \ldots, V_n, \ldots$ a *chain*.

Without loss of generality, we shall assume that there exist ultrafilters t_1^1, t_1^2.

We construct a first chain $V_1, \ldots, V_n, \ldots$, where $V_1 = U_1^1$. We first discuss all the cases that may occur when the chain $V_1, \ldots, V_n, \ldots, V_\ell$ is finite.

Case I. $t_{i_\ell}^{k_\ell} \not\supseteq \bigcup_{m,p} U_p^m$. If $\ell - n + 1$ is even, V_n is labeled 0. All the other sets V_n are labeled 1.

Case II. $U_p^m \in t_{i_\ell}^{k_\ell}$ and U_p^m is adjacent to some V_n. If $\ell - n + 1$ is even, V_n is labeled 0. All other sets V_n are labeled 1.

Case III. $V_j \in t_{i_\ell}^{k_\ell}$ and $\ell - j$ is odd. If n is odd, V_n is labeled 1. All other sets V_n are labeled 0. The important point here is that V_j and V_ℓ are assigned different labels.

Case IV. $V_j \in t_{i_\ell}^{k_\ell}$, $\ell - j$ is even and $V_\ell \neq U_1^2$. Labels are assigned to all V_n except V_ℓ. If $\ell - n$ is odd, V_n is labeled 1. All other V_n are labeled 0.

Case V. $V_j \in t_{i_\ell}^{k_\ell}$, $\ell - j$ is even and $V_\ell = U_1^2$. Labels are assigned to all V_n except $V_{\ell-1}, V_\ell$. If $\ell - n$ is odd, V_n is labeled 0. All other V_n are labeled 1.

Case VI. $U_p^m \in t_{i_\ell}^{k_\ell}$ and U_p^m is a labeled set.[32] If $\ell - n + 1$ is even, V_n is assigned the same label as U_p^m. All other V_n are assigned the other label (recall that the only possible labels are 0 and 1).

Case VII. $U_p^m \in t_{i_\ell}^{k_\ell}$ and the set adjacent to U_p^m is labeled.[33] The procedure is exactly the same as in Case II.

If none of the above seven cases occur, there remains only

Case VIII. The chain $V_1, \ldots, V_n, \ldots$ is infinite. If n is odd, V_n is labeled 1. All other V_n are labeled 0.

If in Cases I, II, III, VI, VII, VIII there exists a labeled set $U_{i(k)}^k$ for each k, let U denote the union of all sets labeled 1; then it is easy to see that $U \notin \mathcal{A}_k$ for all k. Otherwise, if U_1^2 is unlabeled, define $V^* = U_1^2$; if U_1^2 is labeled, take the minimal k^* such that $U_1^{k^*}, U_2^{k^*}$ are unlabeled sets and define $V^* = U_1^{k^*}$. In Case IV, V^* is defined as the set adjacent to V_ℓ. In Case V, V^* is defined as the set adjacent to $V_{\ell-1}$.

If the above construction does not produce the set U, construct a second chain $V^* = V_1, \ldots, V_n, \ldots$, a third, etc. It is possible that Cases IV and V will alternate infinitely many times; after that (if possible) we define V^* as in Case I. The construction will end when there exists a labeled set $U_{i(k)}^k$ for each k (it is clear that two adjacent sets cannot both be labeled). The union of all sets labeled 1 will be a set $U \notin \mathcal{A}_k$ for all k. $\square$

Remark 8.2. The condition that the algebras in the first part of Theorem III will be σ-algebras is obviously meaningful only when one is considering an infinite sequence of algebras (see Corollary 2.1 and Remark 2.1).

---

[32] In the construction of the first chain, U_p^m can be labeled only with the digit 1.

[33] Case VII never occurs when the first chain is constructed.

Remark 8.3. It is obvious that the required set U in the first part of Theorem III has the form $\bigcup_i U_{\beta_i}^{\alpha_i} (\alpha_1 < \alpha_2 < \ldots < \alpha_i < \ldots)$. This is the refinement of the first part of Theorem III promised at the end of Section 2.

Proof of the first part of Theorem IV. With each set U_i^k we associate an $\mathcal{A}_k$-special ultra-filter $s_i^k \ni U_i^k$ or $\mathcal{A}_k$-similar ultrafilters s_i^k, t_i^k; $U_i^k \in s_i^k$, $U_i^k \notin t_i^k$. In the latter case there are three possibilities:

(α) $U_p^m \in t_i^k$;

(β) case (α) does not occur, but there exists a sequence $U_{p_1}^{m_1}, \ldots, U_{p_r}^{m_r}, \ldots, m_1 < m_2 < \ldots < m_r < \ldots$, such that $t_i^k \supseteq \bigcup_r U_{p_r}^{m_r}$;

(γ) neither case (α) nor (β) occurs.

If $k > 2$ we shall refer to the sets $U_1^k, \ldots, U_{n_k}^k$ as adjacent. We are now going to label some of the sets U_i^k with digits 0,1,2. If at least one of the ultrafilters $s_1^k, \ldots, s_{n_k}^k$ is $\mathcal{A}_k$-special, choose one set $U_{i_0}^k$ such that $s_{i_0}^k$ is an $\mathcal{A}_k$-special ultrafilter and label this set 1. We shall assume that there are no sets adjacent to $U_{i_0}^k$, i.e., any adjacent set is not of the form U_i^k. If there exists a labeled set $U_{i(k)}^k$ for each k, let U denote the union of all labeled sets; obviously, $U \notin \mathcal{A}_k$ for all k. Otherwise, we shall consider chains. By a *chain* we mean a finite or infinite sequence of sets U_i^k, satisfying conditions (1), (2), (3) in the proof of the first part of Theorem III, and the following condition instead of the previous condition (4):

(4*) V_n does not have an adjacent set labeled 0 or 1, but it may have adjacent sets labeled 2.

Without loss of generality, we shall assume that there exist ultrafilters t_1^1, t_2^1.

We construct the first chain $V_1, \ldots, V_n, \ldots$, where $V_1 = U_1^1$. We first list all the cases in which the chain $V_1, \ldots, V_n, \ldots, V_\ell$ is finite.

Case I*. Case (γ) holds for $t_{i_\ell}^{k_\ell}$. Proceed as in Case I in the proof of the first part of Theorem III.

Cases II*, III*, IV*, V*. These cases are exactly the same as Cases II, III, IV, V in the proof of the first part of Theorem III.

Case VI*. $U_p^m \in t_{i_\ell}^{k_\ell}$ and U_p^m is a labeled set. If U_p^m is labeled 0 or 2, proceed as in Case VI in the proof of the first part of Theorem III for U_p^m labeled 0. If U_p^m is labeled 1, proceed as in Case VI in the proof of the first part of Theorem III for U_p^m labeled 1.[34]

Case VII*. $U_p^m \in t_{i_\ell}^{k_\ell}$, U_p^m is an unlabeled set, and a set adjacent to U_p^m is labeled 0 or 1.[35] Proceed as in Case II in the proof of the first part of Theorem III.

The next possibility is entirely new:

Case I**. Case (β) holds for $t_{i_\ell}^{k_\ell}$. There are two subcases:

Subcase(a**): There exists a sequence $U_{p_1}^{m_1}, \ldots, U_{p_r}^{m_r}, \ldots$ all of whose sets are assigned the same label (0 or 1) and $t_{i_\ell}^{k_\ell} \ni \bigcup_r U_{p_r}^{m_r}$. If $\ell - n + 1$ is even, assign V_n the same label. All other sets V_n are labeled differently (using only labels 0 and 1).[36]

Subcase(b**): There exists a sequence $U_{p_1}^{m_1}, \ldots, U_{p_r}^{m_r}, \ldots$ of sets, not labeled 0 or 1, such that $t_{i_\ell}^{k_\ell} \ni \bigcup_r U_{p_r}^{m_r}$; among the sets $U_1^{m_r}, \ldots, U_{n_{m_r}}^{m_r}$ there are three unlabeled sets (possibly including $U_{p_r}^{m_r}$) and none of the sets $V_1, \ldots, V_\ell$ occur. If $\ell - n + 1$ is even, label V_n with 0. All other sets V_n are labeled 1. All sets $U_{p_r}^{m_r}$ are labeled 2.[37]

If none of the above cases is effective, there remains only Case VIII* which is exactly the same as Case VIII in the proof of the first part of Theorem III.

If in Cases I*, II*, III*, VI*, VII*, I**, VIII* there exists for every k a set $U_{i(k)}^k$ labeled 0 or 1, let U denote the union of all sets labeled 1; then it is easy to see that $U \notin \mathcal{A}_k$ for all k.[38] Otherwise, if U_1^2 is unlabeled, define $V^* = U_1^2$; if U_1^2 is labeled (it cannot be labeled 2), let k^* be the minimum number such that none of the sets $U_1^{k^*}, \ldots, U_{n_{k^*}}^{k^*}$ is labeled 0 or 1 (some of them may be labeled 2 and two of them are unlabeled), and let V^* denote one of the unlabeled ones. In Case IV*, pick any unlabeled set adjacent to V_ℓ and denote it by V^*. In Case V*, pick any unlabeled set adjacent to $V_{\ell-1}$ and denote it by V^*.

If the above construction has not produced a set U, construct a second chain $V^* = V_1, \ldots, V_n, \ldots$, a third, and so on. Cases IV* and V* may alternate infinitely many times;

---

[34] In the construction of the first chain, U_p^m can be labeled only 1.

[35] In the construction of the first chain, Case VII* cannot occur.

[36] In the construction of the first chain, subcase (a**) may occur, but all the sets $U_{p_r}^{m_r}$ are labeled 1.

[37] Only in subcase (b**) we label some of the sets U_i^k with the digit 2.

[38] In the construction of the first chain in subcase (b**), we cannot construct the set U, since $U_{p_r}^{m_r}$ does not have an adjacent set labeled 0 or 1.

after that (if possible) define a set V^* as in Case I*. Our construction will end when there exists for every k a set $U^k_{i(k)}$ labeled either 0 or 1 (it is obvious that two adjacent sets cannot be assigned these labels). Denote the union of all the sets labeled 1 by U. Clearly, $U \notin \mathcal{A}_k$ for all k. □

Remark 8.4. It is obvious that the required set U in the first part of Theorem IV has the form $\bigcup_i U^{\alpha_i}_{\beta_i}$ ($\alpha_i < \alpha_2 < \ldots < \alpha_i < \ldots$). This is the refinement of the first part of Theorem IV promised at the end of Section 2.

Example 8.1. Let $|X| = \aleph_0$ and let n be a natural number. Arrange the elements of X in a matrix, each element appearing only once:

$$\begin{pmatrix} a_1^1 & \cdots & a_n^1 \\ \cdots\cdots\cdots \\ a_1^k & \cdots & a_n^k \\ \cdots\cdots\cdots \end{pmatrix}.$$

For every $i \leq n$, consider the set $X_i = \{a_i^1, \ldots, a_i^k, \ldots\}$, and let $a_i \in \beta X_i \backslash X_i$. Identify all the compact sets $\beta X_1, \ldots, \beta X_n$ in the natural way, so that $a_1, \ldots, a_n$ are identified. We define a ω-saturated algebra $\mathcal{A}_k$ for each k, as follows: $\ker \mathcal{A}_k$ is the union of k $\mathcal{A}_k$-similar sets:

$$\{a_1^k, a_1\}, \ldots, \{a_n^k, a_n\}.$$

Obviously, $\mathcal{A}_k$ is not a σ-algebra and there exists no $U \notin \mathcal{A}_k$ for all k.

This example implies the truth of the second part of Theorem IV.

9. THE INVERSE PROBLEM

Let $\mathfrak{A} = \{\mathcal{A}_1, \ldots, \mathcal{A}_k, \ldots\}$ be an at most countable family of algebras. In this section, on the assumption that there exists no set not a member of all the algebras $\mathcal{A}_k$, we shall see what conclusions can be drawn about the algebras $\mathcal{A}_k$. To that end we now define the *rank* $\mathfrak{r}(\mathfrak{A})$ *of a sequence* $\mathfrak{A}$.

Definition 9.1. If there exists a set not a member of any algebra in $\mathfrak{A}$, then $\mathfrak{r}(\mathfrak{A})=0$. Otherwise, $\mathfrak{r}(\mathfrak{A})$ is the least number (a natural number or $\aleph_0$) with the property: $\mathfrak{A}$ contains a family $\mathfrak{B}$ of $\mathfrak{r}(\mathfrak{A})$ algebras such that every subset of X is a member of at least one algebra in $\mathfrak{B}$.

If $\mathcal{A}_k \neq \mathcal{P}(X)$ for all k, $\mathfrak{r}(\mathfrak{A})$ may be any natural number (except for 1 and 2), 0 or $\aleph_0$. It is obvious that $\mathfrak{r}(\mathfrak{A}) = 1$ if and only if $\mathcal{A}_k = \mathcal{P}(X)$ for some k.

Theorem 9.1. *(1) If $0 < \mathfrak{r}(\mathfrak{A}) < \aleph_0$, then $\mathfrak{A}$ contains a family $\mathfrak{B}$ of ω-saturated algebras such that $|\mathfrak{B}| = \mathfrak{r}(\mathfrak{A}) = \mathfrak{r}(\mathfrak{B})$. (2) If $\mathfrak{r}(\mathfrak{A}) = \aleph_0$ and all algebras in $\mathfrak{A}$ are almost σ-algebras, then $\mathfrak{A}$ contians $\aleph_0$ ω-saturated algebras and at least one strictly simple algebra.*

Proof. (1) Let $0 < n = \mathfrak{r}(\mathfrak{A}) < \aleph_0$, and let

$$\mathfrak{B} = \{\mathcal{A}_1, \ldots, \mathcal{A}_n\}, \mathfrak{r}(\mathfrak{B}) = n.$$

Suppose that $\mathcal{A}_n$ is not ω-saturated, i.e., there exist $\aleph_0$ pairwise disjoint sets, each of which is not in $\mathcal{A}_n$. If $n = 1$, then $\mathcal{A}_n = \mathcal{P}(X)$. But this contradicts the assumption that $\mathcal{A}_n \neq \mathcal{P}(X)$. If $n > 1$, there exists D such that $D \notin \mathcal{A}_k$ for all $k < n$. By Theorem 4.1 (and also from the proof of Proposition 3.1 which is considerably easier than the proof of Theorem 4.1), there exists a set Q such that $Q \notin \mathcal{A}_k$ for all $k \leq n$. This is again a contradiction.

(2) We first prove that $\mathfrak{A}$ contains $\aleph_0$ ω-saturated algebras. By Theorem 5.3, we may assume that all the $\mathcal{A}_k$'s are simple almost σ-algebras. Suppose that all the algebras $\mathcal{A}_1, \ldots, \mathcal{A}_{k_0}$ are ω-saturated, but if $k > k_0$, then $\mathcal{A}_k$ is strictly simple. It is easy to construct a matrix

$$\begin{pmatrix} U_1^{k_0+1} & & \\ U_1^{k_0+2} & U_2^{k_0+2} & \\ \cdots\cdots\cdots\cdots\cdots & \\ U_1^k & \cdots & U_{k-k_0}^k \\ \cdots\cdots\cdots\cdots\cdots & \end{pmatrix}$$

of pairwise disjoint sets and a set W such that $W \notin \mathcal{A}_k$ if $k \leq k_0$, $U_i^k \notin \mathcal{A}_k$ $(k > k_0)$, $W \cap \bigcup_{k,i} U_i^k = \emptyset$, and

$$\overline{\bigcup_{k,i} U_i^k} \cap \ker \mathcal{A}_m = \emptyset$$

if $m \leq k_0$. By Theorem 5.1, we can construct a set U such that $W \cup U \notin \mathcal{A}_k$ for all k, a contradiction.

We now prove that $\mathfrak{A}$ contains at least one strictly simple algebra. Again by Theorem 5.3, we may assume that all the $\mathcal{A}_k$'s are simple almost σ-algebras. Suppose, moreover, that they are all ω-saturated. For every finite sequence $\mathcal{A}_1, \ldots, \mathcal{A}_k$, consider the appropriate sets S_k, T_k.[39] Let

$$S_k = \{s_1^k, \ldots, s_k^k\},$$

where s_i^k is either an $\mathcal{A}_i$-special ultrafilter or s_i^k has an $\mathcal{A}_i$-similar ultrafilter t_i^k. The set T_k is the set of all t_i^k. Of course, $S_k \cap T_k = \emptyset$. The fact that the set $\ker \mathcal{A}_1$ is finite implies that there is an infinite sequence

$$N^1 = \{n_1^1 < n_2^1 < \ldots < n_m^1 < \ldots\}$$

such that

$$s_1^{n_1^1} = s_1^{n_2^1} = \ldots = s_1^{n_m^1} = \ldots$$

and either each of the ultrafilters $s_1^{n_m^1}$ is $\mathcal{A}_1$-special, or for each m one has an appropriate ultrafilter $t_1^{n_m^1}$ and

$$t_1^{n_1^1} = t_1^{n_2^1} = \ldots = t_1^{n_m^1} = \ldots.$$

The fact that the set $\ker \mathcal{A}_2$ is finite implies that there is an infinite subsequence of N^1,

$$N^2 = \{n_1^2 < n_2^2 < \ldots < n_m^2 < \ldots\}$$

such that

$$s_2^{n_1^2} = s_2^{n_2^2} = \ldots = s_2^{n_m^2} = \ldots$$

and either each of the ultrafilters $s_2^{n_m^2}$ is $\mathcal{A}_2$-special, or for each m one has an appropriate ultrafilter $t_2^{n_m^2}$ and

$$t_2^{n_1^2} = t_2^{n_2^2} = \ldots = t_2^{n_m^2} = \ldots.$$

The construction continues with sequences $N^3, N^4, \ldots$. By construction

$$N^1 \supset N^2 \supset \ldots \supset N^k \supset \ldots.$$

---

[39] Note that here, in contrast to the case in the proof of Theorem 4.1, we cannot indicate rules for constructing the sets S_{k+1}, T_{k+1}.

Let

$$S = \left\{ s_1^{n_1^1}, s_2^{n_1^2}, \ldots, s_k^{n_1^k}, \ldots \right\},$$

and let T be the collection of all the ultrafilters $t_k^{n_1^k}$. It is easy to verify that $S \cap T = \emptyset$. Since all points of $S \cup T$ are irregular, there exists $Q \subset X$ such that $S \subset \overline{Q}$, $\overline{Q} \cap T = \emptyset$. Clearly, $Q \notin \mathcal{A}_k$ for all k. This contradiction shows that $\mathfrak{A}$ must contain a strictly simple algebra. $\square$

Having proved that $\mathfrak{A}$ contains at least one strictly simple algebra, we have incidentally proved the following

Proposition 9.1. *Let $\mathcal{A}_1, \ldots, \mathcal{A}_k, \ldots$ be a countable sequence of ω-saturated σ-algebras, and for every k there exists a set Q_k such that $Q_k \notin \mathcal{A}_i$ if $i \leq k$. Then there exists $Q \notin \mathcal{A}_k$ for all k.*

Example 9.1. There exists a countable family of σ-algebras $\mathfrak{A} = \{\mathcal{A}_0, \mathcal{A}_{n,i}^j\}$ such that $\mathcal{A}_0$ is strictly simple, all the algebras $\mathcal{A}_{n,i}^j$ are ω-saturated, $\mathfrak{r}(\mathfrak{A}) = \aleph_0$, $\mathfrak{r}(\{\mathcal{A}_{n,i}^j\}) = 0$. Each of these algebras will contain X.

Construction. Consider a matrix of different irregular points of βX :

$$\begin{pmatrix} a_1^1 & & \\ a_1^2 & a_2^2 & \\ \cdots\cdots\cdots & \\ a_1^n & \cdots & a_{2^n}^n \\ \cdots\cdots\cdots & \end{pmatrix}.$$

To each point a_i^n we associate two algebras $\mathcal{A}_{n,i}^1, \mathcal{A}_{n,i}^2$ with

$$\ker \mathcal{A}_{n,i}^1 = \{a_i^n, a_{2i-1}^{n+1}\}, \quad \ker \mathcal{A}_{n,i}^2 = \{a_i^n, a_{2i}^{n+1}\}.$$

Define

$$\mathrm{sp}\mathcal{A}_0 = \{a_i^n \mid n \text{ is odd}\},$$

and suppose that any two distinct ultrafilters in $\mathrm{sp}\mathcal{A}_0$ are $\mathcal{A}_0$-similar.

If Q is not a member of any algebra of the form $\mathcal{A}_{n,i}^j$, then either $a_i^n \ni Q$ for all odd n and $a_i^n \not\ni Q$ for all even n, or $a_i^n \ni Q$ for all even n and $a_i^n \not\ni Q$ for all odd n. But any such set Q is a member of $\mathcal{A}_0$. $\square$

If all points a_i^n are contained in $\beta X \setminus X$ (this is possible if $|X|$ is a σ-measurable cardinal), then each algebra in $\mathfrak{A}$ is the algebra of all measurable sets of a two-valued measure.

In light of Proposition 9.1, there is some importance to the following

Remark 9.1. Clearly, for every k there exists a set Q_k such that $Q_k \notin \mathcal{A}_0$, $\mathcal{A}_{n,i}^j$; $1 \le n \le k$. But there exists no set which is not a member of any algebra in $\mathfrak{A}$. This is because $\mathcal{A}_0$ is not an ω-saturated algebra, though all the other algebras are ω-saturated. Recall that all the algebras in $\mathfrak{A}$ are σ-algebras.

Remark 9.2. At the end of Section 8 we constructed Example 8.1 of a countable sequence of algebras $\mathfrak{A} = \{\mathcal{A}_1, \ldots, \mathcal{A}_k, \ldots\}$ such that each $\mathcal{A}_k$ is ω-saturated and $\mathfrak{r}(\mathfrak{A}) = \aleph_0$. In this example none of the algebra $\mathcal{A}_k$ is a σ-algebra.

10. Finite Sequences of Algebras (2)

In this section Theorems V, VII, IX, and XI are proved.

The proof of the first part of Theorem V for $n = 1, 2$ is obvious. To prove the first part of the theorem in the general case we need the following

Theorem 10.1. *Consider a finite sequence of algebras $\mathcal{A}_1, \ldots, \mathcal{A}_n, \mathcal{A}_{n+1}$. Suppose there exist pairwise disjoint sets $V, U_1, \ldots, U_n$ such that if $U_k \subset Q$, $V \cap Q = \emptyset$, then $Q \notin \mathcal{A}_k$, $1 \le k \le n$. Moreover, assume that there exist more than $\frac{5n}{2}$ pairwise disjoint sets not members of $\mathcal{A}_{n+1}$. Then there exist pairwise disjoint sets $V', U_1', \ldots, U_n', U_{n+1}'$ such that if $U_k' \subset Q$, $V' \cap Q = \emptyset$, then $Q \notin \mathcal{A}_k$, $1 \le k \le n+1$.*

Proof. For each k, $1 \le k \le n$, we consider either an $\mathcal{A}_n$-special ultrafilter s_k or a pair of $\mathcal{A}_k$-similar ultrafilters s_k, t_k, such that in either case $U_k \in s_k$ and in the second case also $V \in t_k$. Define

$$S_n = \{s_1, \ldots, s_n\},$$

and let T_n be the set of all t_k. Let

$$K = \{s_i \in S_n \mid \text{ either there exists } j \ne i \text{ such that } t_i = t_j,$$
$$\text{or } s_i \text{ is an } \mathcal{A}_i\text{-special ultrafilter}\}.$$

It is clear that

$$|S_n \cup T_n| \le \frac{3|K|}{2} + 2(n - |K|).$$

By assumption

$$|\ker \mathcal{A}_{n+1}| > \frac{5n}{2}.$$

Thus there exists $L \subset \ker \mathcal{A}_{n+1}$ such that

$$L \cap (S_n \cup T_n) = \emptyset$$

and

$$|L| > \frac{5n}{2} - \frac{3|K|}{2} - 2n + 2|K| = \frac{(n + |K|)}{2} \ge |K|.$$

If L contains an $\mathcal{A}_{n+1}$-special ultrafilter, denote it by s_{n+1}. Suppose now that L does not contain $\mathcal{A}_{n+1}$-special ultrafilters. If for every ultrafilter in L there is an $\mathcal{A}_{n+1}$-similar ultrafilter in K, then L contains two $\mathcal{A}_{n+1}$-similar ultrafilters (recall that $(|L| > |K|)$). Denote one of them by s_{n+1} and the other by t_{n+1}. Now consider the case in which there exist $\mathcal{A}_{n+1}$-similar ultrafilters q, r such that $q \in L$, $r \notin K$. If $r \notin S_n$, we put $s_{n+1} = q$, $t_{n+1} = r$; but if $r \in S_n \setminus K$, say $r = s_{i_0}$, we put $s_{n+1} = q$, $t_{i_0} = t_{n+1} = r$, and denote the ultrafilter previously denoted by t_{i_0} by s_{i_0}. In sum: we have constructed a set of pairwise distinct ultrafilters

$$S_{n+1} = \{s_1, \ldots, s_n, s_{n+1}\}.$$

Let T_{n+1} be the set of all t_k, where k can assume values from 1 to $n + 1$. It is important that

$$S_{n+1} \cap T_{n+1} = \emptyset.$$

The construction of the required sets $V', U_1', \ldots, U_n', U_{n+1}'$ is now a simple matter. $\square$

Remark 10.1. Consider a countable sequence of algebras $\mathcal{A}_1, \ldots, \mathcal{A}_k, \ldots$ satisfying all the conditions of Theorem VI except one: they are not necessarily σ-algebras. Along the lines of the proof of Theorem 10.1, we can inductively construct sets

$$S_1, T_1; S_2, T_2; \ldots; S_k, T_k; \ldots .$$

Let $a = s_{i_0} \in S_k$, $b = t_{i_0} \in T_k$, but in the construction of S_{k+1}, T_{k+1} we have $s_{i_0} = b$, $t_{i_0} = a$. Then in the construction of S_ℓ, T_ℓ, $\ell > k$, we have: $s_{i_0} = b$, $t_{i_0} = a$. It follows from the previous arguments that we can construct a countable set of pairwise distinct ultrafilters

$$S = \{s_1, \ldots, s_k, \ldots\}.$$

If s_k is not an $\mathcal{A}_k$-special ultrafilter, we consider $\mathcal{A}_k$-similar ultrafilters s_k, t_k. Define T as the set of all ultrafilters t_k. Then

$$S \cap T = \emptyset, \quad S \cup T = \bigcup_k (S_k \cup T_k).$$

However, we cannot state that there are sets $V, U_1, \ldots, U_k, \ldots$ as required in the statement of Theorem VI.

The next theorem refines the first part of Theorem V:

Theorem 10.2. *Let $\mathcal{A}_1, \ldots, \mathcal{A}_n$ be a finite sequence of algebras. Suppose that for each k, $1 \le k \le n$, we are given a finite sequence of pairwise disjoint sets $Q_1^k, \ldots, Q_{m_k}^k \notin \mathcal{A}_k$, where $m_k > \frac{5}{2}(k-1)$ if $k \ne 2$. Suppose, moreover, that if $n > 1$ there exist three pairwise disjoint sets $Q_{\beta_1}^{\alpha_1}, Q_{\beta_2}^{\alpha_2}, Q_{\beta_3}^{\alpha_3}$, where each of the superscripts α_i is either 1 or 2 and two of them are different. Then for each set Q_i^k one can construct a set $\hat{Q}_i^k \subset Q_i^k$ such that $\hat{Q}_i^k \notin \mathcal{A}_k$; either $\hat{Q}_i^k = \hat{Q}_j^\ell$ or $\hat{Q}_i^k \cap \hat{Q}_j^\ell = \emptyset$; and there exist pairwise disjoint sets $V, \hat{Q}_{i_1}^{\gamma_1}, \ldots, \hat{Q}_{i_n}^{\gamma_n}$ such that if $\hat{Q}_{i_k}^{\gamma_k} \subset Q, V \cap Q = \emptyset$, then $Q \notin \mathcal{A}_k$, $1 \le k \le n$.*

Proof. To each set Q_i^k we associate either an $\mathcal{A}_k$-special ultrafilter s_i^k or a pair of $\mathcal{A}_k$-similar ultrafilters s_i^k, t_i^k; $s_i^k \ni Q_i^k$, $t_i^k \not\ni Q_i^k$. Our goal is to construct corresponding sets S_n and T_n such that S_n is a set of pairwise distinct ultrafilters

$$\{s_1, \ldots, s_n\}$$

and

$$S_n \subset \bigcup_{k,i} \{s_i^k\}.$$

The possibility of constructing such sets S_n, T_n obviously implies the assertion of the theorem. The construction will proceed by induction: we shall successively construct sets

$$S_1, T_1; S_2, T_2; \ldots; S_p, T_p; \ldots; S_n, T_n.$$

This will be done in such a way that $|S_p| = p$ and

$$S_p \subset \bigcup_{k \leq p, i} \{s_i^k\}.$$

The construction of S_1, T_1 is obvious.

Let $n = 2$. Suppose that (see the assumptions of the theorem)

$$s_1^1 = s_{\beta_1}^{\alpha_1}, \ s_1^2 = s_{\beta_2}^{\alpha_2}, \ s_2^2 = s_{\beta_3}^{\alpha_3}.$$

We shall construct S_2, T_2 by examining the various possibilities. There are two basic cases:

Case (1). Either s_1^1 is an $\mathcal{A}_1$-special ultrafilter or $t_1^1 \notin \{s_1^2, s_2^2\}$;

Case (2). $t_1 \in \{s_1^2, s_2^2\}$.

Let $1 < p < n$ and suppose that S_p and T_p have already been constructed. We proceed to construct S_{p+1} and T_{p+1}, using the same reasoning as in the proof of Theorem 10.1. Let $S_p = \{s_1, \dots, s_p\}$. We define a set $K^* \subset S_p$ as follows: an ultrafilter s_i will be in K^* if and only if it satisfies either of two conditions:

(a) either there exists $j \neq i$ such that $t_i = t_j$, $j \leq p$, or s_i is an $\mathcal{A}_i$-special ultrafilter;

(b) s_i is not an $\mathcal{A}_i$ special ultrafilter; $t_i \neq t_j$ whenever $i \neq j$, $j \leq p$; t_i is not any of the ultrafilters s_ℓ^{p+1}.

On the strength of the arguments which are almost exactly the same as in the proof of Theorem 10.1, there exists a set

$$L \subset \{s_1^{p+1}, \dots, s_{m_{p+1}}^{p+1}\}$$

such that

(1) $L \cap (S_p \cup T_p) = \emptyset$;

(2) $|L| > |K^*|$.

The rest of the proof is an almost literal repetition of the end of the proof of Theorem 10.1. Let L contain an $\mathcal{A}_{p+1}$-special ultrafilter, denote it by s_{p+1}. Suppose now that L does not contain $\mathcal{A}_{p+1}$-special ultrafilters. If for every ultrafilter in L there is an $\mathcal{A}_{p+1}$-similar ultrafilter in K^*, then L contains two $\mathcal{A}_{p+1}$-similar ultrafilters (recall that $|L| > |K^*|$). Denote one of them by s_{p+1} and the other by t_{p+1}. Now consider the case in which there exist $\mathcal{A}_{p+1}$-similar ultrafilters q, r such that $q \in L$, $r \notin K^*$. If $r \notin S_p$, we put $s_{p+1} = q$, $t_{p+1} = r$. If $r \in S_p \backslash K^*$, say $r = s_{i_0}$, then t_{i_0} is equal to s_ℓ^{p+1}. We put $s_{p+1} = q$, $t_{i_0} = t_{p+1} = r$, and denote the ultrafilter previously denoted by t_{i_0} by s_{i_0}. In sum: we have constructed a set of pairwise distinct ultrafilters

$$S_{p+1} = \{s_1 \ldots, s_p, s_{p+1}\}$$

and

$$S_{p+1} \subset \bigcup_{k \leq p+1, i} \{s_i^k\}.$$

The set T_{p+1} is defined in the natural way. $\square$

Remark 10.2. Suppose that the sequence of algebras in Theorem 10.2 is not finite but countable: $\mathcal{A}_1, \ldots, \mathcal{A}_k, \ldots$. As usual, we consider for each set Q_i^k either a $\mathcal{A}_k$-special ultrafilter s_i^k or $\mathcal{A}_k$-similar ultrafilters s_i^k, t_i^k. Analysis of the proof of the theorem now implies that we can construct sets S, T as in Remark 10.1 such that also $S \subset \bigcup_{k,i} \{s_i^k\}$.

Proof of the second part of Theorem V.

Let $n = 1$. Let a_1, q_1 be $\mathcal{A}_1$-similar ultrafilters, $\ker \mathcal{A}_1 = \{a_1, q_1\}$, and define an algebra $\mathcal{A}_2^* = \mathcal{A}_1$. It is obvious that the algebras $\mathcal{A}_1, \mathcal{A}_2^*$ satisfy conditions of the second part of Theorem V for $n = 1$.

Let $n = 2$. Consider algebras $\mathcal{A}_2, \mathcal{A}_3^*$ defined as follows ($\mathcal{A}_1$ is the same algebra as before); a_2, q_1 are $\mathcal{A}_2$-similar ultrafilters, and

$$\ker \mathcal{A}_2 = \{a_2, q_1\}, \quad \ker \mathcal{A}_3^* = \{a_1, a_2, b_1, b_2, q_1\};$$

$\{a_1, b_1\}$ and $\{a_2, b_2\}$ are $\mathcal{A}_3^*$-similar sets, and q_1 is an $\mathcal{A}_3^*$-special ultrafilter.

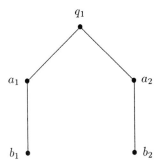

It is obvious that the algebras $\mathcal{A}_1, \mathcal{A}_2, \mathcal{A}_3^*$ satisfy conditions of the second part of Theorem V for $n = 2$.

Let $n = 3$. Consider algebras $\mathcal{A}_3, \mathcal{A}_4^*$ defined as follows. The $\mathcal{A}_3$-similar sets are:

$$\{a_1, b_1\}, \ \{a_2, b_2\}, \ \{a_3, q_2\}.$$

The $\mathcal{A}_4^*$-similar sets are:

$$\{a_1, b_1\}, \ \{a_2, b_2\}, \ \{a_3, q_1, q_2\}.$$

Note that there exist neither $\mathcal{A}_3$-special nor $\mathcal{A}_4^*$-special ultrafilters. In general, in the constructions given here all the algebras except $\mathcal{A}_3^*$ contain X.

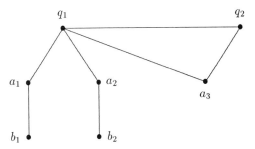

The algebras $\mathcal{A}_1, \mathcal{A}_2, \mathcal{A}_3, \mathcal{A}_4^*$ satisfy conditions of the second part of Theorem V for $n = 3$.

Let $n = 4$. We now define algebras $\mathcal{A}_4, \mathcal{A}_5^*$. The $\mathcal{A}_4$-similar sets are:

$$\{a_1, b_1\}, \ \{a_2, b_2\}, \ \{a_3, q_1\}, \ \{a_4, q_2\}.$$

The $\mathcal{A}_5^*$-similar sets are:

$$\{a_1, b_1\}, \ \{a_2, b_2\}, \ \{a_3, b_3\}, \ \{a_4, b_4\}, \ \{q_1, q_2\}.$$

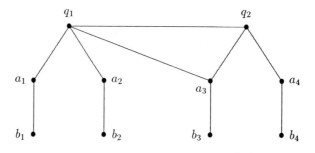

The algebras $\mathcal{A}_1, \mathcal{A}_2, \mathcal{A}_3, \mathcal{A}_4, \mathcal{A}_5^*$ satisfy conditions of the second part of Theorem V for $n = 4$.

Let $n = 5$. Now let $\mathcal{A}_5 = \mathcal{A}_6^*$. The $\mathcal{A}_5$-similar sets are:

$$\{a_1, b_1\}, \ \{a_2, b_2\}, \ \{a_3, b_3\}, \ \{a_4, b_4\}, \ \{q_1, q_2\}, \ \{a_5, q_3\}.$$

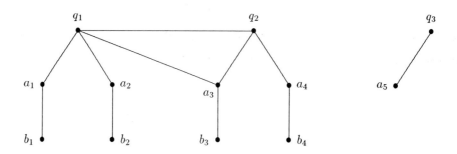

The algebras $\mathcal{A}_1, \mathcal{A}_2, \mathcal{A}_3, \mathcal{A}_4, \mathcal{A}_5, \mathcal{A}_6^*$ satisfy conditions of the second part of Theorem V for $n = 5$.

Let $n = 6$. Now define algebras $\mathcal{A}_6, \mathcal{A}_7^*$ as follows. The $\mathcal{A}_6$-similar sets are:

$$\{a_1, b_1\}, \ \{a_2, b_2\}, \ \{a_3, b_3\}, \ \{a_4, b_4\}, \ \{a_5, q_1, q_2\}, \ \{a_6, q_3\}.$$

The $\mathcal{A}_7^*$-similar sets are:

$$\{a_1, b_1\}, \ \{a_2, b_2\}, \ \{a_3, b_3\}, \ \{a_4, b_4\}, \ \{a_5, b_5\}, \ \{a_6, b_6\}, \ \{q_1, q_2, q_3\}.$$

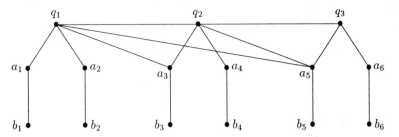

The algebras $\mathcal{A}_1, \mathcal{A}_2, \mathcal{A}_3, \mathcal{A}_4, \mathcal{A}_5, \mathcal{A}_6, \mathcal{A}_7^*$ satisfy conditions of the second part of Theorem V for $n = 6$.

Continuing in this way for every natural $n > 6$, we can construct algebras $\mathcal{A}_1, \ldots, \mathcal{A}_n, \mathcal{A}_{n+1}^*$ such that: $\mathcal{A}_1, \ldots, \mathcal{A}_n, \mathcal{A}_{n+1}^*$ satisfy the assumptions of the second part of Theorem V. $\square$

To prove the first part of Theorem VII we shall need the following

Theorem 10.3. *Consider a finite sequence of algebras $\mathcal{A}_1, \ldots, \mathcal{A}_n, \mathcal{A}_{n+1}$. Suppose there exist pairwise disjoint sets $U_1, \ldots, U_n, V_1, \ldots, V_n$ such that if $U_k \subset Q$, $V_k \cap Q = \emptyset$, then $Q \notin \mathcal{A}_k$, $1 \leq k \leq n$. Moreover, assume that there exist more than $4n$ pairwise disjoint sets not members of $\mathcal{A}_{n+1}$. Then there exist pairwise disjoint sets $U_1', \ldots, U_n', U_{n+1}', V_1', \ldots, V_n', V_{n+1}'$ such that if $U_k' \subset Q$, $V_k' \cap Q = \emptyset$, then $Q \notin \mathcal{A}_k$, $1 \leq k \leq n+1$.*

Proof. For every k, $1 \leq k \leq n$, we consider either an $\mathcal{A}_k$-special ultrafilter s_k or a pair of $\mathcal{A}_k$-similar ultrafilters s_k, t_k, such that in either case $U_k \in s_k$ and in the second case also $V_k \in t_k$. Define

$$S_n^0 = \{s_1, \ldots, s_n\},$$

and let T_n^0 denote the set of all t_k. Since $|S_n^0 \cup T_n^0| \leq 2n$ and $\ker |\mathcal{A}_{n+1}| > 4n$, there exists $L \subset \ker \mathcal{A}_{n+1} \backslash (S_n^0 \cup T_n^0)$ such that $|L| > 2n$. If L contains an $\mathcal{A}_{n+1}$-special ultrafilter, we denote it by s_{n+1}. If for every ultrafilter in L there is an $\mathcal{A}_{n+1}$-similar ultrafilter in $S_n^0 \cup T_n^0$, then L contains two $\mathcal{A}_{n+1}$-similar ultrafilters. Denote one of them by s_{n+1}, the other by t_{n+1}. There remains only one possibility: L does not contain $\mathcal{A}_{n+1}$-special ultrafilters and there exist $\mathcal{A}_{n+1}$-similar ultrafilters s_{n+1}, t_{n+1}; $s_{n+1} \in L$, $t_{n+1} \notin (S_n^0 \cup T_n^0)$. In all cases we can construct pairwise disjoint sets $U_1', \ldots, U_n', U_{n+1}', V_1', \ldots, V_n', V_{n+1}'$ such that $U_k' \in s_k$ for all $k \leq n+1$. If t_k exists, then also $V_k' \in t_k$, but if there is no t_k (i.e. s_k is an $\mathcal{A}_k$-special ultrafilter), then $V_k' = \emptyset$. $\square$

Remark 10.3. Let $\mathcal{A}_1, \ldots, \mathcal{A}_k, \ldots$ be a countable sequence of algebras satisfying all the assumptions of Theorem VIII except one: they are not necessarily σ-algebras. Following the proof of Theorem 10.3, we can construct a set of pairwise distinct ultrafilters

$$S^0 = \{s_1, \ldots, s_k, \ldots\}.$$

If s_k is not an $\mathcal{A}_k$-special ultrafilter, we consider $\mathcal{A}_k$-similar ultrafilters s_k, t_k. It is not possible that $t_{k_1} = t_{k_2}$ but $k_1 \neq k_2$. The set T^0 is the set of all t_k, and $S^0 \cap T^0 = \emptyset$. But we cannot ensure the existence of sets $U_1, \ldots, U_k, \ldots, V_1, \ldots, V_k, \ldots$ as required in Theorem VIII.

Analysis of the proof of Theorem 10.3 implies the following refined version of the first part of Theorem VII:

Theorem 10.4. *Let $\mathcal{A}_1, \ldots, \mathcal{A}_n$ be a finite sequence of algebras. Suppose that for each k, $1 \leq k \leq n$, we are given a finite sequence of pairwise disjoint sets $Q_1^k, \ldots, Q_{m_k}^k \notin \mathcal{A}_k$, where $m_k > 4(k-1)$. Then for each set Q_i^k one can construct a set $\hat{Q}_i^k \subset Q_i^k$ such that $\hat{Q}_i^k \notin \mathcal{A}_k$; either $\hat{Q}_i^k = \hat{Q}_j^\ell$ or $\hat{Q}_i^k \cap \hat{Q}_j^\ell = \emptyset$; there exist pairwise disjoint sets $\hat{Q}_{i_1}^1, \ldots, \hat{Q}_{i_n}^n$, $V_1, \ldots, V_n$ such that if $\hat{Q}_{i_k}^k \subset Q$, $V_k \cap Q = \emptyset$, then $Q \notin \mathcal{A}_k$, $1 \leq k \leq n$.*

Remark 10.4. Suppose that the sequence of algebras in Theorem 10.4 is not finite but countable: $\mathcal{A}_1, \ldots, \mathcal{A}_k, \ldots$. As usual, we consider for each set Q_i^k either an $\mathcal{A}_k$-special ultrafilter s_i^k or $\mathcal{A}_k$-similar ultrafilters s_i^k, t_i^k. Analysis of the proof of this theorem now implies that we can construct sets S^0, T^0 as in Remark 10.3 such that also $s_k \in \{s_1^k, \ldots, s_{m_k}^k\}$.

Proof of the second part of Theorem VII. For every natural number n we can construct algebras $\mathcal{A}_1, \ldots, \mathcal{A}_n, \mathcal{A}_{n+1}$ such that $|\ker \mathcal{A}_k| = 4(k-1) + 2$, $1 \leq k \leq n$, $|\ker \mathcal{A}_{n+1}| = 4n$, and there exist no corresponding sets $U_1, \ldots, U_n, U_{n+1}, V_1, \ldots, V_n, V_{n+1}$. It may be assumed that X is a member of all these algebras. To illustrate, we describe the kernels of these algebras in the case $n = 4$:

$\mathcal{A}_1$-similar sets: $\{a_1, b_1\}$;

$\mathcal{A}_2$-similar sets: $\{a_1, c_1\}, \{b_1, c_2\}, \{a_2, b_2\}$;

$\mathcal{A}_3$-similar sets: $\{a_1, c_1\}, \{b_1, c_2\}, \{a_2, c_3\}, \{b_2, c_4\}, \{a_3, b_3\}$;

$\mathcal{A}_4$-similar sets: $\{a_1, c_1\}, \{b_1, c_2\}, \{a_2, c_3\}, \{b_2, c_4\}, \{a_3, c_5\}, \{b_3, c_6\}, \{a_4, b_4\}$;

$\mathcal{A}_5$-similar sets: $\quad \{a_1,c_1\}, \{b_1,c_2\}, \{a_2,c_3\}, \{b_2,c_4\}, \{a_3,c_5\}, \{b_3,c_6\}, \{a_4,c_7\}, \{b_4,c_8\}.$

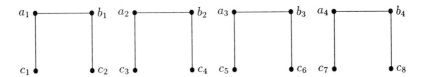

By analogy, such algebras can be constructed for any $n > 4$. $\square$

Before proving Theorem IX, we shall prove the following

Theorem 10.5. *(1) Consider a finite sequence of algebras $\mathcal{A}_1,\ldots,\mathcal{A}_n,\mathcal{A}_{n+1}$. Suppose there exist pairwise disjoint sets $V,U_1,\ldots,U_n$ such that if $U_k \subset Q$, $V \cap Q = \emptyset$, then $Q \notin \mathcal{A}_k$, $1 \le k \le n$. Moreover, assume that there exist pairwise disjoint sets $U_1^*,\ldots,U_\ell^* \notin \mathcal{A}_{n+1}$ such that $\ell > \frac{3n}{2}$ and*

$$\bigcup_{k=1}^{n} U_k \cap \bigcup_{k=1}^{\ell} U_k^* = \emptyset.$$

Then there exist pairwise disjoint sets $V',U_1',\ldots,U_n',U_{n+1}'$ such that

$$\left(\bigcup_{k=1}^{n} U_k \cup \bigcup_{k=1}^{\ell} U_k^* \right) \supset \bigcup_{k=1}^{n+1} U_k',$$

and if $U_k' \subset Q, V' \cap Q = \emptyset$, then $Q \notin \mathcal{A}_k$, $1 \le k \le n+1$. (2) The bound $\ell > \frac{3n}{2}$ is best possible in the following sense: For any natural number n one can construct algebras $\mathcal{A}_1,\ldots,\mathcal{A}_n,\mathcal{A}_{n+1}$ and sets $V,U_1,\ldots,U_n,U_1^*,\ldots,U_\ell^*$ satisfying all the assumptions of the first part of our theorem except one: $\ell = [\frac{3n}{2}]$. But there do not exist pairwise disjoint sets $V',U_1',\ldots,U_n',U_{n+1}'$ such that if $U_k' \subset Q, V' \cap Q = \emptyset$, then $Q \notin \mathcal{A}_k$, $1 \le k \le n+1$.

Proof of the first part of Theorem 10.5. As in the proof of Theorem 10.1, we consider sets of ultrafilters S_n and T_n. To each U_k^* there corresponds either an $\mathcal{A}_{n+1}$-special ultrafilter s_k^* or a pair of $\mathcal{A}_{n+1}$-similar ultrafilters s_k^*, t_k^* such that $U_k^* \in s_k$, and if t_k^* exists, then also $U_k^* \notin t_k$. Define

$\tilde{S} = \{s_k \in S_n \mid t_k \text{ exists and there exist } s_i, t_i \text{ such that } t_k = t_i, \ k \ne i\},$

$|\tilde{S}| = m,$

$$\tilde{T} = \{t_k \in T_n \mid s_k \in \tilde{S}\},$$

$$S^* = \{s_1^*, \dots, s_\ell^*\},$$

$$S' = \{s_k \in S_n \backslash \tilde{S} \mid \text{ either } t_k \text{ exists and } t_k \notin S^*, \text{ or } t_k \text{ does not exist}\},$$

$$|S'| = q,$$

$$\hat{S} = S^* \backslash (S_n \cup T_n),$$

$$\hat{T} = \{t_k^* \mid s_k^* \in \hat{S}\}.$$

It is clear that

$$|S^* \cap (S_n \cup T_n)| = |S^* \cap T_n| = |S^* \cap \tilde{T}| + |S^* \cap (T_n \backslash \tilde{T})| \leq \frac{m}{2} + n - m - q = n - \frac{m}{2} - q.$$

Therefore,

$$|\hat{S}| > \frac{3n}{2} - (n - \frac{m}{2} - q) \geq m + q.$$

Henceforth we shall consider ultrafilters s_i^* only in $\hat{S}$, and ultrafilters t_i^* only in $\hat{T}$. If there is an $\mathcal{A}_{n+1}$-special ultrafilter s_i^*, we put $s_{n+1} = s_i^*$. Suppose now that for each s_i^* there is a t_i^*. The following cases are possible:

(1) there exists $t_{i_0}^* \notin S_n$;

(2) all the t_i^* are in $\tilde{S} \cup S'$;

(3) there exists $t_{i_0}^* = s_{k_0} \in S_n \backslash (\tilde{S} \cup S')$.

In case (1) we define $s_{n+1} = s_{i_0}^*$, $t_{n+1} = t_{i_0}^*$. In case (2) there exists $\mathcal{A}_{n+1}$-similar ultrafilters $s_{k_1}^*, s_{k_2}^*$, since $|\tilde{S} \cup S'| = m + q$, but $|\hat{S}| > m + q$; so we put $s_{n+1} = s_{k_1}^*$, $t_{n+1} = s_{k_2}^*$. In case (3) we denote s_{k_0} by t_{k_0} and t_{k_0} by s_{k_0}; $s_{n+1} = s_{i_0}^*$, $t_{n+1} = t_{i_0}^*$. It is obvious that $s_{k_0} \in S^* \backslash \hat{S}$. Therefore $s_{k_0} \neq s_{n+1}$ ($s_{n+1} \in \hat{S}$).

In all cases, we have constructed a set of pairwise distinct ultrafilters

$$S_{n+1} = \{s_1, \dots, s_n, s_{n+1}\};$$

T_{n+1} is the set of all the t_k's, $1 \leq k \leq n+1$ (it is possible for some k, t_k's not to exist). Obviously, $S_{n+1} \cap T_{n+1} = \emptyset$. Moreover,

$$S_{n+1} \subset \overline{\bigcup_{k=1}^{n} U_k} \cup \overline{\bigcup_{k=1}^{\ell} U_k^*}.$$

The construction of the required sets $V', U_1', \dots, U_n', U_{n+1}'$ is now a simple matter. $\square$

Proof of Theorem IX. Let $n > 1$. Suppose that suitable sets S_k, T_k can be constructed for the algebras $\mathcal{A}_1, \ldots, \mathcal{A}_k$, where $k < n$, and moreover

$$S_k \subset \overline{\bigcup_{i \leq k, j} U_j^i}.$$

(If $n = 1$, the construction of S_1, T_1 is obvious, and $S_1 \subset \overline{\bigcup_j U_j^1}$.)Considering the set $U_1^{k+1}, \ldots, U_{m_{k+1}}^{k+1}$ and using the first part of Theorem 10.5, we construct S_{k+1}, T_{k+1}, where

$$S_{k+1} \subset \overline{\bigcup_{i \leq k+1, j} U_j^i}.$$

Thus the required sets S_n and T_n can be constructed by induction. $\square$

Remark 10.5. It is easy to show that in Theorem IX one may choose the sets $U_1, \ldots, U_n$ in such a way that

(1) $U_k \subset U_{i_k}^{p_k}$;

(2) if $k_1 \neq k_2$, then $U_{i_{k_1}}^{p_{k_1}} \neq U_{i_{k_2}}^{p_{k_2}}$;

(3) if $p_{k_0} \neq p_k$ for all $k \neq k_0$, then $p_{k_0} = k_0$;

(4) if $p_{k_1} = p_{k_2}$ $(k_1 \neq k_2)$, then $p_k \neq p_{k_1}$ for $k \notin \{k_1, k_2\}$; if $k_1 < k_2$, then $p_{k_1} = p_{k_2} = k_2$.

This is the refinement of Theorem IX promised at the end of Section 2.

Remark 10.6. Let $\mathcal{A}_1, \ldots, \mathcal{A}_k, \ldots$ be a countable sequence of algebras satisfying all the assumptions of Theorem X except one: they are not necessarily σ-algebras. By analogy with Remark 10.1, we can construct suitable sets S and T, but the assertion of Theorem X does not follow.

Now we proceed to the

Proof of the second part of Theorem 10.5.

If $n = 1$, $\mathcal{A}_1 = \mathcal{A}_2$, $\ker \mathcal{A}_1 = \{a_1, b_1\}$, and a_1, b_1 are $\mathcal{A}_1$-similar ultrafilters.

If $n = 2$, $\ker \mathcal{A}_1 = \{a_1, b_1\}$, $\ker \mathcal{A}_2 = \{a_2, b_1\}$, $\ker \mathcal{A}_3 = \{a_1, a_2, b_1, b_2, b_3\}$. At the same time: a_1, b_1 are $\mathcal{A}_1$-similar ultrafilters; a_2, b_1 are $\mathcal{A}_2$-similar ultrafilters; b_1 is an $\mathcal{A}_3$-special ultrafilter, and $\{a_1, b_2\}, \{a_2, b_3\}$ are $\mathcal{A}_3$-similar sets.

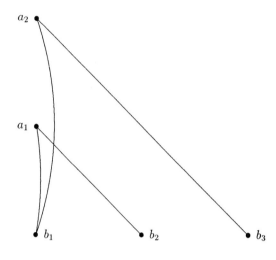

If $n \geq 3$, then $X \in \mathcal{A}_1, \ldots, \mathcal{A}_n, \mathcal{A}_{n+1}$.

Suppose first that *n is odd, $n \geq 3$.* Then

$$\ker \mathcal{A}_k = \{a_k, b_{\left[\frac{k+1}{2}\right]}\}$$

if $k \leq n$. We now exhibit all $\mathcal{A}_{n+1}$-similar sets:

$$\{b_1, \ldots, b_{\frac{n+1}{2}}\},$$
$$\{a_1, b_{\frac{n+3}{2}}\},$$
$$\cdots\cdots\cdots\cdots$$
$$\{a_m, b_{\frac{n+2m+1}{2}}\},$$
$$\cdots\cdots\cdots\cdots$$
$$\{a_{n-1}, b_{\left[\frac{3n}{2}\right]}\}.$$

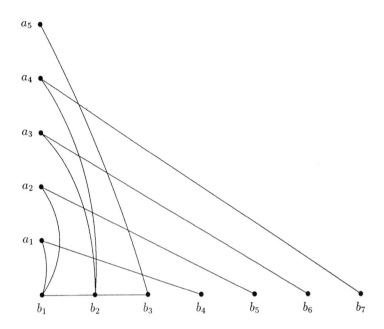

Suppose now that *n is even*, $n \geq 4$. Then

$$\ker \mathcal{A}_k = \left\{a_k, b_{\left[\frac{k+1}{2}\right]}\right\}$$

if $k \leq n$. We now exhibit all $\mathcal{A}_{n+1}$-similar sets:

$$\{b_1, \ldots, b_{\frac{n}{2}}\},$$
$$\{a_1, b_{\frac{n+2}{2}}\},$$
$$\cdots\cdots\cdots\cdots$$
$$\{a_m, b_{\frac{n+2m}{2}}\},$$
$$\cdots\cdots\cdots\cdots$$
$$\{a_n, b_{\frac{3n}{2}}\}.$$

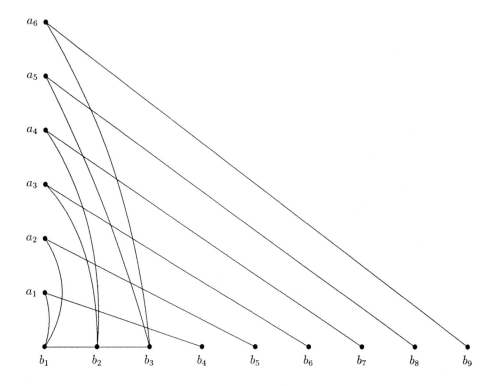

The sets

$$V, U_1, \ldots, U_n, U_1^*, \ldots, U_\ell^*$$

must satisfy the following conditions:

$$V \in b_1, \ldots, b_{[\frac{n+1}{2}]};$$

$$U_k \in a_k, \ 1 \le k \le n;$$

$$U_k^* \in b_k, \ 1 \le k \le \ell = \left[\frac{3n}{2}\right].$$

It is easy to check that corresponding

$$V', U_1', \ldots, U_n', U_{n+1}'$$

do not exist. □

Using by now familiar arguments we easily prove the following

Theorem 10.6. *Consider a finite sequence of algebras* $\mathcal{A}_1, \ldots, \mathcal{A}_n, \mathcal{A}_{n+1}$. *Suppose there exist pairwise disjoint sets* $U_1, \ldots, U_n, V_1, \ldots, V_n$ *such that if* $U_k \subset Q$, $V_k \cap Q = \emptyset$, *then* $Q \notin \mathcal{A}_k$, $1 \le k \le n$. *Moreover, assume that there exist pairwise disjoint sets* $U_1^*, \ldots, U_\ell^* \notin \mathcal{A}_{n+1}$ *such that* $\ell > 3n$ *and* $\bigcup_{k=1}^n U_k \cap \bigcup_{k=1}^\ell U_k^* = \emptyset$. *Then there exist pairwise disjoint sets* $U_1', \ldots, U_n', U_{n+1}', V_1', \ldots, V_n', V_{n+1}'$ *such that if* $U_k' \subset Q$, $V_k' \cap Q = \emptyset$, *then* $Q \notin \mathcal{A}_k$, $1 \le k \le n+1$.

Incidentally, it is obvious that we can also demand that

$$\left(\bigcup_{k=1}^n U_k \cup \bigcup_{k=1}^\ell U_k^* \right) \supset \bigcup_{k=1}^{n+1} U_k'.$$

It is easy to see that the bound $\ell > 3n$ in Theorem 10.6 is best possible. Just as the first part of Theorem 10.5 implies the proof of Theorem IX, Theorem 10.6 implies the proof of Theorem XI.

Remark 10.7. It is easy to show that in Theorem XI one may choose the sets $U_1, \ldots, U_n$ in such a way that $U_k \subset U_{i_k}^k$. This is the refinement of Theorem XI promised at the end of Section 2.

Remark 10.8. Consider a countable sequence of algebras $\mathcal{A}_1, \ldots, \mathcal{A}_k, \ldots$ satisfying all the assumptions of Theorem XII except one: they are not necessarily σ-algebras. By analogy with Remark 10.3, we can construct suitable sets S^0 and T^0. But we cannot ensure the existence of sets $U_1, \ldots, U_k, \ldots, V_1, \ldots, V_k, \ldots$ as required in Theorem XII.

Using our methods, it is rather easy to prove the following theorem, which should be compared with Theorem IX.

Theorem 10.7. *Consider a finite sequence of algebras* $\mathcal{A}_1, \ldots, \mathcal{A}_n$. *Suppose there exists a matrix of pairwise disjoint sets*

$$\begin{pmatrix} U_1^1 & \cdots & U_{m_1}^1 \\ \cdots\cdots\cdots\cdots\cdots \\ U_1^n & \cdots & U_{m_n}^n \end{pmatrix},$$

where $m_k > 2(k-1)$, $U_i^k \notin \mathcal{A}_k$. *Then there exist pairwise disjoint sets* $V, U_1, \ldots, U_n$ *such that if* $U_k \subset Q$, $V \cap Q = \emptyset$, *then* $Q \notin \mathcal{A}_k$, $1 \le k \le n$.

11. Countable Sequences of Algebras (2)

In this section we shall prove the following

Theorem 11.1. *Consider two countable sequences of σ-algebras*

$$\mathfrak{A} = \{\mathcal{A}_1, \ldots, \mathcal{A}_k, \ldots\}, \ \mathfrak{B} = \{\mathcal{B}_1, \ldots, \mathcal{B}_k, \ldots\}.$$

All the algebras in $\mathfrak{A}$ are simple; none of those in $\mathfrak{B}$ are simple. Then there exist pairwise disjoint sets $W, U_1, \ldots, U_k, \ldots, V_1, \ldots, V_k, \ldots$ such that

(1) $\ker \mathcal{A}_k \subset \overline{W}$;

(2) if $U_k \subset Q$, $V_k \cap Q = \emptyset$, then $Q \notin \mathcal{B}_k$.[40]

The proof of Theorem 11.1 requires several lemmas and definitions.

Definition 11.1. An algebra $\mathcal{A}$ is said to be absolutely inseparable if there exists an outer inseparable set $M \subset \ker \mathcal{A}$ such that two different ultrafilters in M are $\mathcal{A}$-similar ultrafilters.

Lemma 11.1. *Consider a σ-algebra $\mathcal{A}$ which is not absolutely inseparable; let M be an outer separable set, and*

$$K = \{x \in \ker \mathcal{A} \mid x \text{ has an } \mathcal{A}\text{-similar ultrafilter } y \in \overline{M}\}.$$

Then K is an outer separable set.

Proof. Let

$$M^* = \{y \in \overline{M} \mid \text{ there exist } \mathcal{A}\text{-similar ultrafilters } x, y\}.$$

If $M^* = \emptyset$, there is nothing to prove. Let $M^* \neq \emptyset$, $Z \subset \beta X$, $|Z| \leq \aleph_0$, $\overline{Z} \supset M^*$,

$$Z^* = \{z \in Z \mid z \text{ is not an } \mathcal{A}\text{-special ultrafilter}\}.$$

Clearly $\overline{Z^*} \supset M^*$. Let

$$Z^* = \{z_1, \ldots, z_n, \ldots\},$$

$$C_n = \{x \in \ker \mathcal{A} \mid x, z_n \text{ are } \mathcal{A}\text{-similar ultrafilters}\} \cup \{z_n\}.$$

---

[40]It is obvious that Theorem 11.1 is a generalization of Theorem 5.3.

Let $G \subset X$, $\overline{G} \cap K \neq \emptyset$. Then $\overline{G} \cap C_n \neq \emptyset$ for some n. Suppose the contrary. Then for every n there exists $W_n \in \mathcal{A}$ such that $C_n \subset \overline{W}_n$, $G \cap W_n = \emptyset$. Since $\mathcal{A}$ is a σ-algebra, $\mathcal{A} \ni \bigcup_n W_n$. But there exist $\mathcal{A}$-similar ultrafilters x, y such that $G \in x$; $y \in M^* \subset \overline{\bigcup_n W_n}$. Thus $\mathcal{A} \not\ni \bigcup_n W_n$, a contradiction. Hence $\overline{G} \cap C_n \neq \emptyset$ for some n. Since $\mathcal{A}$ is not absolutely inseparable, it follows that C_n is outer separable. Thus K is an outer separable set. $\square$

Definition 11.2. Consider an algebras $\mathcal{A}$. We call ultrafilters a, b an $\mathcal{A}$-inseparability pair if one of the following two condtions holds:

(1) $a = b$, and for any $G \in a$ the set $\overline{G} \cap \ker \mathcal{A}$ is outer inseparable and the set

$$\{y \in \beta X \backslash \overline{G} \mid y \text{ has an } \mathcal{A}\text{-similar ultrafilter } x \ni G\}$$

    is outer separable;

(2) $a \neq b$, and for any $G_1 \in a$, $G_2 \in b$ the sets $\overline{G}_1 \cap \ker \mathcal{A}$ and

$$\{y \in \overline{G}_2 \mid y \text{ has an } \mathcal{A}\text{-similar ultrafilter } x \ni G_1\}$$

    are outer inseparable.

Lemma 11.2. *If $\mathcal{A}$ is an inseparable σ-algebra, there exists an $\mathcal{A}$-inseparability pair.*

Proof. Let c be an inseparability point of $\ker \mathcal{A}$. If c, c is not an $\mathcal{A}$-inseparability pair then, since βX is compact, there exist $a, b \in \beta X$ such that $a \neq b$ and if $G_1 \in a$, $G_2 \in b$, then the set

$$\{y \in \overline{G}_2 \mid y \text{ has an } \mathcal{A}\text{-similar ultrafilter } x \ni G_1\}$$

is outer inseparable. If $\mathcal{A}$ is not absolutely inseparable, then by Lemma 11.1, a, b is an $\mathcal{A}$-inseparability pair. But if $\mathcal{A}$ is absolutely inseparable, the existence of an $\mathcal{A}$-inseparability pair is obvious. $\square$

Lemma 11.3. *Consider two countable sequences of σ-algebras*

$$\mathfrak{A} = \{\mathcal{A}_1, \ldots, \mathcal{A}_k, \ldots\}, \quad \mathfrak{B} = \{\mathcal{B}_1, \ldots, \mathcal{B}_k, \ldots\}.$$

All the algebras in $\mathfrak{A}$ are separable; all those in $\mathfrak{B}$ are inseparable. Then there exist pairwise disjoint sets $W, U_1, \ldots, U_k, \ldots, V_1, \ldots, V_k, \ldots$ such that

(1) $\ker \mathcal{A}_k \subset \overline{W}$;

(2) if $U_k \subset Q$, $V_k \cap Q = \emptyset$, then $Q \notin \mathcal{B}_k$.[41]

Proof. For every $\mathcal{B}_k$, consider a $\mathcal{B}_k$-inseparability pair b_1^k, b_2^k. Let $Z \subset \beta X$, $|Z| \leq \aleph_0$, and $b_1^k, b_2^k \in Z$, $\ker \mathcal{A}_k \subset \overline{Z}$ for all k. We can construct pairwise disjoint sets

$$U_1^1, \ldots, U_4^1, V_1^1, \ldots, V_4^1$$

such that if $V_i^1 = \emptyset$, then U_i^1 is a member of a $\mathcal{B}_1$-special ultrafilter, but if $V_i^1 \neq \emptyset$, then there exist two $\mathcal{B}_1$-similar ultrafilters, one of which contains U_i^1 and the other contains V_i^1. At the same time

$$Z \cap \overline{\bigcup_{i=1}^{4}(U_i^1 \cup V_i^1)} = \emptyset.$$

If $\mathcal{B}_1$ is not absolutely inseparable, these sets can be constructed by means of Lemma 11.1. But if $\mathcal{B}_1$ is absolutely inseparable, the construction is obvious. Continuing, we now construct sets

$$U_1^2, \ldots, U_9^2, V_1^2, \ldots, V_9^2$$

such that if $V_i^2 = \emptyset$, then U_i^2 is a member of a $\mathcal{B}_2$-special ultrafilter, while if $V_i^2 \neq \emptyset$, then there exist two $\mathcal{B}_2$-similar ultrafilters one of which contains U_i^2 and the other V_i^2. At the same time

$$Z \cap \overline{\bigcup_{i=1}^{9}(U_i^2 \cup V_i^2)} = \emptyset,$$

and the sets

$$U_1^1, \ldots, U_4^1, V_1^1, \ldots, V_4^1$$
$$U_1^2, \ldots, U_9^2, V_1^2, \ldots, V_9^2$$

are pairwise disjoint. Continuing in this way, we get a matrix of pairwise disjoint sets

$$\begin{pmatrix} U_1^1 & \cdots & U_4^1 & V_1^1 & \cdots & V_4^1 \\ U_1^2 & \cdots & U_9^2 & V_1^2 & \cdots & V_9^2 \\ \cdots\cdots\cdots\cdots\cdots\cdots\cdots\cdots\cdots\cdots \\ U_1^k & \cdots & U_{(k+1)^2}^k & V_1^k & \cdots & V_{(k+1)^2}^k \\ \cdots\cdots\cdots\cdots\cdots\cdots\cdots\cdots\cdots\cdots \end{pmatrix}.$$

[41] It is obvious that Lemma 11.3 is a generalization of Theorem 5.2.

Applying Lemma 5.1 twice,[42] we choose sets

$$U^1_{i_1}, V^1_{i_1}; U^2_{i_2}, V^2_{i_2}; \ldots; U^k_{i_k}, V^k_{i_k}; \ldots$$

such that

$$Z \cap \overline{\bigcup_k (U^k_{i_k} \cup V^k_{i_k})} = \emptyset.$$

Throughout this construction: if $V^k_{i_k} = \emptyset$, then $U^k_{i_k}$ is a member of a $\mathcal{B}_k$-special ultrafilter; but if $V^k_{i_k} \neq \emptyset$, then there exist two $\mathcal{B}_k$-similar ultrafilters, one of which contains $U^k_{i_k}$ and the other $V^k_{i_k}$. It remains to define $U_k = U^k_{i_k}$, $V_k = V^k_{i_k}$, and take some $W \subset X$ such that $Z \subset \overline{W}$ and

$$W \cap \bigcup_k (U_k \cup V_k) = \emptyset. \quad \square$$

Definition 11.3. A separable algebra $\mathcal{A}$ is said to be absolutely separable if there exists $Q \subset X$ such that

(1) $\overline{Q} \cap \ker \mathcal{A} \neq \emptyset$;

(2) all points of $\overline{Q} \cap \ker \mathcal{A}$ are regular;

(3) any two different ultrafilters in $\overline{Q} \cap \ker \mathcal{A}$ are $\mathcal{A}$-similar.

Lemma 11.4. *Consider a separable σ-algebra $\mathcal{A}$; let $Z \subset \beta X$, $|Z| \leq \aleph_0$, and suppose that all points of Z are irregular. Let $Q \subset X$, $\overline{Q} \cap Z = \emptyset$, and suppose that all points of $\overline{Q} \cap \ker \mathcal{A}$ are regular. Finally, suppose there exist $\mathcal{A}$-similar ultrafilters q, r such that $Q \in q$, $r \in \overline{Z}$. Then $\mathcal{A}$ is absolutely separable.*

Proof. Let Q' be a subset of Q such that $Q' \in q$ and the set $\overline{Q'} \cap \ker \mathcal{A}$ contains no $\mathcal{A}$-special ultrafilters.

Case I. $r \in Z$. Let $M \subset \overline{Q'}$, $|M| = \aleph_0$, $\overline{M} \supset \overline{Q'} \cap \ker \mathcal{A}$,

$$M' = \{x \in M \mid x, r \text{ are not } \mathcal{A}\text{-similar ultrafilters}\}.$$

It is a simple matter to prove that $q \notin \overline{M'}$. Let $Q'' \subset Q'$, $Q'' \in q$ and $\overline{Q''} \cap M' = \emptyset$. If $x \in \overline{Q''} \cap \ker \mathcal{A}$ and $x \neq q$, then x, q are $\mathcal{A}$-similar ultrafilters.

---

[42] Lemma 5.1 is used in accordance with Remark 5.1.

Case II. $r \in \overline{Z} \backslash Z$. It is easy to establish the existence of a $z \in Z$ which has an $\mathcal{A}$-similar ultrafilter in $\overline{Q}$ – but this is exactly Case I. □

Let $\mathcal{A}$ be a σ-algebra; $K, L \subset X$, $K \cap L = \emptyset$. Assume that the set

$$P = \{x \in \overline{L} \mid x \text{ has an } \mathcal{A}\text{-similar ultrafilter } y \ni K\}$$

is not empty, and $P \subset \beta X \backslash X$. Define $\mu_{\mathcal{A}}^{K,L}(M) = 0$ iff $\overline{M} \cap P = \emptyset$, $M \subset X$. It is not hard to prove the following

Lemma 11.5. $\mu_{\mathcal{A}}^{K,L}$ *defines a two-valued measure on* X.

It is obvious that $\mu_{\mathcal{A}}^{K,L}(L) = 1$, and if $\mu_{\mathcal{A}}^{K,L}(M) \neq \emptyset$, then there exist $\mathcal{A}$-similar ultrafilters q, r such that $M \in q$, $K \in r$.

The proof of Theorem 11.1 will make use of the following lemma, whose proof resembles that of Lemma 11.4:

Lemma 11.6. *Let* $K, L \subset X$; $K \cap L = \emptyset$. *Let* $\mathcal{A}$ *be a separable* σ-*algebra and let* x, y *be* $\mathcal{A}$-*similar ultrafilters,* $K \in x$, $L \in y$. *Then there exists* $D \subset K$ *such that* $D \in x$ *and any ultrafilter in* $\overline{D} \cap \ker \mathcal{A}$ *has an* $\mathcal{A}$-*similar ultrafilter in* $\overline{L}$.

Let $\mathcal{A}$ be an almost σ-algebra, $U \subset X$, $\overline{U} \cap \ker \mathcal{A} \neq \emptyset$. As in Section 5, we define a σ-additive measure $\mu_{\mathcal{A}}^{U}$ on X as follows:

$$\mu_{\mathcal{A}}^{U}(M) = 0 \text{ iff } \overline{M \cap U} \cap \ker \mathcal{A} = \emptyset, \ \mu_{\mathcal{A}}^{U}(M) = 1 \text{ iff } \mu_{\mathcal{A}}^{U}(X \backslash M) = 0.$$

If $(\overline{U} \cap \ker \mathcal{A}) \cap X = \emptyset$, then $\mu_{\mathcal{A}}^{U}$ is a two-valued measure.

Proof of Theorem 11.1. By Lemma 11.3, we may assume that the algebras in $\mathfrak{B}$ are strictly separable. Let Z be the union of all spectra of $\mathfrak{A}$ and $\mathfrak{B}$. To each algebra $\mathcal{B}_k$ we associate a set $Q_k \subset X$ satisfying the following conditions:

(1) $\overline{Q}_k \cap Z = \emptyset$;

(2) $\overline{Q}_k \cap \ker \mathcal{B}_k \neq \emptyset$ (it is obvious that all points of $\overline{Q}_k \cap \ker \mathcal{B}_k$ are regular);

(3) if $\mathcal{B}_k$ is absolutely separable, then any two different ultrafilters in $\overline{Q}_k \cap \ker \mathcal{B}_k$ are $\mathcal{B}_k$-similar;

(4) the ultrafilters in $\overline{Q}_k \cap \ker \mathcal{B}_k$ are either all $\mathcal{B}_k$-special or all not $\mathcal{B}_k$-special.

Consider the two-valued measures $\left\{ \mu_{\mathcal{B}_k}^{Q_k} \right\}$. Using (if it is necessary) the reasoning from Section 5 and the Gitik-Shelah theorem, we take pairwise disjoint sets

$$Q_1', \ldots, Q_k', \ldots$$

such that $Q_k' \subset Q_k$ and $\mu_{\mathcal{B}_k}^{Q_k}(Q_k') \neq 0$. The set Q_k' contains $\aleph_0$ pairwise disjoint sets each of which is $\mu_{\mathcal{B}_k}^{Q_k}$-nonmeasurable. Therefore, by Lemma 5.1, we may assume that

$$Z \cap \overline{\bigcup_k Q_k'} = \emptyset.$$

Consider a subset of the natural numbers

$$N_1 = \{k \mid k > 1, \text{and there exist } \mathcal{B}_k\text{-similar ultrafilters}$$

$$q_k, r_k; \ Q_k' \in q_k, Q_1' \in r_k\}.$$

Without loss of generality, we shall assume that $N_1 \neq \emptyset$ and put

$$N_1 = \{k_1^1, \ldots, k_n^1, \ldots\}.$$

Consider the sequence of two-valued measures

$$\mu_0^1 = \mu_{\mathcal{B}_1}^{Q_1'}, \mu_1^1 = \mu_{\mathcal{B}_{k_1^1}}^{Q_{k_1^1}', Q_1'}, \ldots, \mu_n^1 = \mu_{\mathcal{B}_{k_n^1}}^{Q_{k_n^1}', Q_1'}, \ldots.$$

Using (if it is necessary) the reasoning from Section 5 and the Gitik-Shelah theorem, we take pairwise disjoint sets

$$F_0^1, F_1^1, \ldots, F_n^1, \ldots \subset Q_1'$$

such that F_n^1 is μ_n^1-nonmeasurable. If none of the ultrafilters in $\overline{Q_1'} \cap \ker \mathcal{B}_1$ is $\mathcal{B}_1$-special, pick $\mathcal{B}_1$-similar ultrafilters q_1, r_1 such that $F_0^1 \in q_1$. By Lemma 8.1, we may assume that only the following cases are possible:

  (a) $F_0^1 \in r_1$;

  (b) $F_{n_0}^1 \in r_1$, $n_0 > 0$;

  (c) $r_1 \not\supset \bigcup_n F_n^1$.

If $\mathcal{B}_1$ is absolutely separable, we consider case (a). In case (a) we take a set $\hat{F}_0^1 \subset F_0^1$ such that $\hat{F}_0^1 \in q_1$, $\hat{F}_0^1 \notin r_1$. The set F_0^1 will play no futher part in the proof, so we denote

$\hat{F}_0^1$ by F_0^1. In case (b), we take a set $\hat{F}_{n_0}^1 \subset F_{n_0}^1$ such that $\hat{F}_{n_0}^1$ is $\mu_{n_0}^1$-nonmeasurable and $\hat{F}_{n_0}^1 \notin r_1$. The set $F_{n_0}^1$ will play no further part, so we denote $\hat{F}_{n_0}^1$ by $F_{n_0}^1$. Then, by Lemma 11.6, for every k_n^1 we consider a set $Q''_{k_n^1} \subset Q'_{k_n^1}$ such that $\overline{Q''_{k_n^1}} \cap \ker \mathcal{B}_{k_n^1} \neq \emptyset$, and for any ultrafilter in $\overline{Q''_{k_n^1}} \cap \ker \mathcal{B}_{k_n^1}$ there exists a $\mathcal{B}_{k_n^1}$-similar ultrafilter in $\overline{F_n^1}$. The sets $Q'_{k_n^1}$ will not be needed any more, so we denote $Q''_{k_n^1}$ by $Q'_{k_n^1}$. After introducing the new sets $Q'_{k_n^1}$, we consider case (c) which splits into two subcases:

(c*) $Q'_{k_0} \in r_1$, $k_0 > 1$;

(c**) $(\bigcup_n F_n^1 \cup \bigcup_{k>1} Q'_k) \notin r_1$.

Consider subcase (c*). Take a set $Q_{k_0}^* \subset Q'_{k_0}$ such that $\overline{Q_{k_0}^*} \cap \ker \mathcal{B}_{k_0} \neq \emptyset$ and $Q_{k_0}^* \notin r_1$. The set Q'_{k_0} will not plany any further part, so we denote $Q_{k_0}^*$ by Q'_{k_0}.

Thus, in all cases, whenever q_1, r_1 are considered,

$$F_0^1 \in q_1,$$
$$\left(\bigcup_n F_n^1 \cup \bigcup_{k>1} Q'_k\right) \notin r_1,$$

and by Lemma 11.4, $r_1 \notin \overline{Z}$.

Consider a subset of the natural numbers

$$N_2 = \{k \notin N_1 \mid k > 2, \text{and there exist } \mathcal{B}_k\text{-similar ultrafilters}$$
$$q_k, r_k; \ Q'_k \in q_k, Q'_2 \in r_k\}.$$

Without loss of generality, we shall assume that $N_2 \neq \emptyset$ and put

$$N_2 = \{k_1^2, \ldots, k_n^2, \ldots\}.$$

Consider the sequence of two-valued measures

$$\mu_0^2 = \mu_{\mathcal{B}_2}^{Q'_2}, \ \mu_1^2 = \mu_{\mathcal{B}_{k_1^2}}^{Q'_{k_1^2}, Q'_2}, \ldots, \mu_n^2 = \mu_{\mathcal{B}_{k_n^2}}^{Q'_{k_n^2}, Q'_2}, \ldots.$$

As before, we consider pairwise disjoint sets

$$F_0^2, F_1^2, \ldots, F_n^2, \ldots \subset Q'_2$$

such that F_n^2 is μ_n^2-nonmeasurable.

Let $2 \in N_1$. For every k_n^2, consider a set $Q''_{k_n^2} \subset Q'_{k_n^2}$ such that $\overline{Q''_{k_n^2}} \cap \ker \mathcal{B}_{k_n^2} \neq \emptyset$ and for any ultrafilter in $\overline{Q''_{k_n^2}} \cap \ker \mathcal{B}_{k_n^2}$ there is a $\mathcal{B}_{k_n^2}$-similar ultrafilter in $\overline{F_n^2}$. The sets $Q'_{k_n^2}$ will not be needed again, so we denote $Q''_{k_n^2}$ by $Q'_{k_n^2}$. We now proceed to examine a set N_3, and so on.

Now let $2 \notin N_1$. If none of the ultrafilters in $\overline{Q'_2} \cap \ker \mathcal{B}_2$ are $\mathcal{B}_2$-special, we pick $\mathcal{B}_2$-similar ultrafilters q_2, r_2 such that $F_0^2 \in q_2$. Obviously, $r_2 \not\ni \bigcup_n F_n^1$ ($2 \notin N_1$). As before, we distinguish several cases:

(a) $F_0^2 \in r_2$;

(b) $F_{n_0}^2 \in r_2$, $n_0 > 0$;

(c) $r_2 \not\ni \bigcup_n F_n^2$.

If $\mathcal{B}_2$ is absolutely separable, we consider case (a). As before, we first consider cases (a) and (b), subsequently introducing new sets $Q'_{k_n^2}$, and only then examine case (c) which again splits into two subcases:

(c*) $Q'_{k_0} \in r_2$, $k_0 > 2$;

(c**) $(\bigcup_n F_n^2 \cup \bigcup_{k>2} Q'_k) \notin r_2$.

All further arguments are anologous to our previous ones, and by Lemma 11.4, $r_2 \notin \overline{Z}$. The next stage is to define sets N_3, and so on.

The final result of the entire process is a matrix of pairwise disjoint sets

$$\begin{pmatrix} F_0^1 & F_1^1 & \cdots & F_n^1 & \cdots \\ F_0^2 & F_1^2 & \cdots & F_n^2 & \cdots \\ \cdots\cdots\cdots\cdots\cdots\cdots \\ F_0^k & F_1^k & \cdots & F_n^k & \cdots \\ \cdots\cdots\cdots\cdots\cdots \end{pmatrix}.$$

The k-th row of this matrix contains at least one element F_0^k. For each k, $\overline{F_0^k} \cap \ker \mathcal{B}_k \neq \emptyset$, and by Lemma 8.1, we may assume that only the following cases are possible:

(1) all ultrafilters in $\overline{F_0^k} \cap \ker \mathcal{B}_k$ are $\mathcal{B}_k$-special;

(2) there exists a set $F_{n_k}^{m_k}$, $m_k < k$, $n_k > 0$, such that for every ultrafilter in $\overline{F_0^k} \cap \ker \mathcal{B}_k$ there is a $\mathcal{B}_k$-similar ultrafilter in $\overline{F_{n_k}^{m_k}}$;

(3) there exist $\mathcal{B}_k$-similar ultrafilters q_k, r_k such that $F_0^k \in q_k$; $r_k \not\ni \bigcup_{i,j} F_j^i$.

In case (2) we denote $F_{n_k}^{m_k}$ by F_k. In case (3) we pick a set D_k such that $D_k \in r_k$ and

$$D_k \cap \bigcup_{i,j} F_j^i = \emptyset.$$

Moreover, $F_{k_1} \neq F_{k_2}$ if $k_1 \neq k_2$, $Z \cap \overline{\bigcup_{i,j} F_j^i} = \emptyset$, $Z \cap \overline{D_k} = \emptyset$ $(r_k \notin \overline{Z})$. By Lemma 11.6, for each D_k we can choose $D_k' \subset D_k$ such that $\overline{D_k'} \cap \ker \mathcal{B}_k \neq \emptyset$, and for every ultrafilter in $\overline{D_k'} \cap \ker \mathcal{B}_k$ there is a $\mathcal{B}_k$-similar ultrafilter in $\overline{F_0^k}$. Let N^* be the set of all natural numbers k for which D_k is considered. Consider the two-valued measures $\left\{ \mu_{\mathcal{B}_k}^{D_k'} \right\}_{k \in N^*}$. Using (if it is necessary) the reasoning from Section 5 and the Gitik-Shelah theorem we take pairwise disjoint sets $\{D_k''\}_{k \in N^*}$ such that $D_k'' \subset D_k'$, $\overline{D_k''} \cap \ker \mathcal{B}_k \neq \emptyset$. The set D_k'' contains $\aleph_0$ pairwise disjoint sets each of which is not a member of $\mathcal{B}_k$. Therefore, by Lemma 5.1, we may assume that

$$Z \cap \overline{\bigcup_{k \in N^*} D_k''} = \emptyset.$$

Now set $U_k = F_0^k$. If all the ultrafilters in $\overline{U}_k \cap \ker \mathcal{B}_k$ are $\mathcal{B}_k$-special, we put $V_k = \emptyset$. If F_k is considered, then $V_k = F_k$. If D_k is considered, then $V_k = D_k''$. Clearly

$$Z \cap \overline{\bigcup_k (U_k \cup V_k)} = \emptyset.$$

Now let $W \subset X$, $Z \subset \overline{W}$, $W \cap \bigcup_k (U_k \cup V_k) = \emptyset$. □

We can now proceed to the

Proof of Theorem VIII. By Theorem 11.1, we may assume that all algebras $\mathcal{A}_k$ are simple. Using the same reasoning as Remark 6.2, we consider ω-saturated algebras $\mathcal{A}_k'$ such that $\mathcal{A}_k' > \mathcal{A}_k$ and $|\ker \mathcal{A}_k'| > 4(k-1)$. As in Remark 10.3, construct sets S^0 and T^0 for the algebras $\mathcal{A}_k'$. All the points of $S^0 \cup T^0$ are irregular, so that the existence of suitable sets $U_1, \ldots, U_k, \ldots, V_1, \ldots, V_k, \ldots$ is assured. □

The proof of Theorem VI is analogous.

Proof of Theorem XII. By Theorem 11.1, we may assume that all the $\mathcal{A}_k$'s are simple. Therefore, for each set U_i^k we can find either an $\mathcal{A}_k$-special ultrafilter s_i^k or a pair of $\mathcal{A}_k$-similar ultrafilters s_i^k, t_i^k $(U_i^k \in s_i^k, U_i^k \notin t_i^k)$, and moreover we may assume that can be done in such a way that s_i^k, t_i^k are irregular points. By Remark 10.8, we can construct sets S^0, T^0 all of whose points are irregular, and thus ensure the existence of the required sets $U_1, \ldots, U_k, \ldots, V_1, \ldots, V_k, \ldots$. $\square$

The proof of Theorem X is analogous.

12. IMPROVEMENT OF SOME MAIN RESULTS

Theorem 12.1. *Consider a countable sequence of σ-algebras $\mathcal{A}_1, \ldots, \mathcal{A}_k, \ldots$ and a matrix of pairwise disjoint sets*

$$\begin{pmatrix} U_1^1 & \cdots & U_{m_1}^1 \\ \cdots\cdots\cdots\cdots \\ U_1^k & \cdots & U_{m_k}^k \\ \cdots\cdots\cdots\cdots \end{pmatrix},$$

where $U_i^k \notin \mathcal{A}_k$. Then every U_i^k contains a subset V_i^k such that:

(1) either there is an $\mathcal{A}_k$-special ultrafilter $v_i^k \ni V_i^k$, or there is a pair of $\mathcal{A}_k$-similar ultrafilters v_i^k, w_i^k such that $V_i^k \in v_i^k$;

(2) w_i^k contains either a set $W'_{k,i}$ or a set $W''_{k,i}$, with the following properties:

(i) if $W'_{k,i} \in w_i^k$, then v_i^k, w_i^k are irregular points;

(ii) either $W'_{k,i} = V_p^m$, and then $w_i^k = v_p^m$, or $W'_{k,i} \cap \bigcup_{m,p} V_p^m = \emptyset$;

(iii) either $W'_{k,i} = W'_{m,p}$, and then $w_i^k = w_p^m$, or $W'_{k,i} \cap W'_{m,p} = \emptyset$;

(iv) the family consisting of all the sets V_i^k, $W''_{k,i}$ and the union of all the sets $W'_{m,p}$ not of the form V_i^k is a family of pairwise disjoint sets.

Proof. If not all ultrafilters in $\overline{U_i^k} \cap \ker \mathcal{A}_k$ are $\mathcal{A}_k$-special, we replace U_i^k by a subset, also denoted by U_i^k. For this new U_i^k it is true that $\overline{U_1^k} \cap \ker \mathcal{A}_k \neq \emptyset$ and none of the ultrafilters in $\overline{U_1^k} \cap \ker \mathcal{A}_k$ are $\mathcal{A}_k$-special. If all the ultrafilters of $\overline{U_i^k} \cap \ker \mathcal{A}_k$ are $\mathcal{A}_k$-special, we have the following possible cases:

I. $U_i^k \in s_i^k \in \ker \mathcal{A}_k$, and s_i^k is an irregular point;

II. All points of $\overline{U_i^k} \cap \ker \mathcal{A}_k$ are regular.

If none of the ultrafilters of $\overline{U_i^k} \cap \ker \mathcal{A}_k$ are $\mathcal{A}_k$-special, we have the following possibilities:

III. There exist $\mathcal{A}_k$-similar ultrafilters s_i^k, t_i^k such that $U_i^k \in s_i^k$, and s_i^k, t_i^k are irregular points;

IV. Case III does not hold, but there exist $\mathcal{A}_k$-similar ultrafilters s_i^k, t_i^k such that $U_i^k \in s_i^k$, $U_{k,i}^* \in t_i^k$, and s_i^k is an irregular point (of course, t_i^k is a regular point); either $U_{k,i}^* \subset U_p^m$ or $U_{k,i}^* \cap \bigcup_{m,p} U_p^m = \emptyset$;

V. All points of $\overline{U_i^k} \cap \ker \mathcal{A}_k$ are regular and there exist $\mathcal{A}_k$-similar ultrafilters s_i^k, t_i^k such that $U_i^k \in s_i^k$, and t_i^k is an irregular point;

VI. All points of $\overline{U_i^k} \cap \ker \mathcal{A}_k$ are regular, but Case V does not hold, and there exist $\mathcal{A}_k$-similar ultrafilters s_i^k, t_i^k such that $U_i^k \in s_i^k$, $U_{k,i}^0 \in t_i^k$ (of course, t_i^k is a regular point); either $U_{k,i}^0 \subset U_p^m$ ($U_i^k \neq U_p^m$) or $U_{k,i}^0 \cap \bigcup_{m,p} U_p^m = \emptyset$.

The possibility of choosing $U_{k,i}^0$ in Case VI follows from Lemma 8.1; similarly, the possibility of choosing $U_{k,i}^*$ in Case IV follows from arguments analogous to those used to prove that lemma and Lemma 6.2 (see Remark 8.1).

Replacing U_i^k's by suitable subsets, we may assume that if Case III holds for some $U_{i_0}^{k_0}$ and $U_p^m \in t_{i_0}^{k_0}$, then $t_{i_0}^{k_0} = s_p^m$. Moreover, we may assume that U_p^m satisfies the conditions of Case I, Case III or Case IV. If Case III holds for some $U_{i_0}^{k_0}$, then either $U_p^m \in t_{i_0}^{k_0}$ or $t_{i_0}^{k_0} \not\supset \bigcup_{m,p} U_p^m$. We may also assume that if in Case IV $U_{k,i}^* \subset U_p^m$, then U_p^m satisfies the conditions of Case II, Case V or Case VI. Similarly, we may assume that if in Case VI $U_{k,i}^0 \subset U_p^m$, then U_p^m satisfies the conditions of Case II, Case V or Case VI. Let

$$\Theta = \{t_i^k \mid U_i^k \text{ satisfies the conditions of Case III and } t_i^k \not\supset \bigcup_{m,p} U_p^m\}.$$

We may also assume that $\overline{U_{k,i}^*} \cap \Theta = \emptyset$, $\overline{U_{m,p}^0} \cap \Theta = \emptyset$.

If U_i^k satisfies the conditions of Case II, we consider the two-valued measure $\nu_{k,i} = \mu_{\mathcal{A}_k}^{U_i^k}$. If it satisfies the conditions of Case IV, we define a two-valued measure $\nu_{k,i}^*$ on X as follows: $\nu_{k,i}^*(M) = 0$ iff

$$\overline{M} \cap \{t \in \overline{U_{k,i}^*} \mid s_i^k, t \text{ are } \mathcal{A}_k\text{-similar ultrafilters}\} = \emptyset.$$

If U_i^k satisfies the conditions of Case V, we define a two-valued measure on X by setting $\nu_{k,i}'(M) = 0$ iff

$$\overline{M} \cap \{s \in \overline{U_i^k} \mid s, t_i^k \text{ are } \mathcal{A}_k\text{-similar ultrafilters}\} = \emptyset.$$

If U_i^k satisfies the conditions of Case VI, we define two two-valued measures:

$$\nu_{k,i}^0 = \mu_{\mathcal{A}_k}^{U_i^k}, \quad \nu_{k,i}^{00} = \mu_{\mathcal{A}_k}^{U_i^k, U_{k,i}^0}.$$

Now suppose that U_1^1 satisfies the conditions of Case II. Consider the family of measures $\mathfrak{M}_1^1$ whose elements are:

(1) all measures $\nu_{k,i}^*$ such that $\nu_{k,i}^*(U_1^1) = 1$;

(2) all measures $\nu_{k,i}^{00}$ such that $\nu_{k,i}^{00}(U_1^1) = 1$;

(3) $\nu_{1,1}$.

Using (if it is necessary) the reasoning from Section 5 and the Gitik-Shelah theorem, we take pairwise disjoint sets

$$L_1^{1,1}, \ldots, L_n^{1,1}, \ldots \subset U_1^1,$$

which can be put in one-to-one correspondence with the measures in $\mathfrak{M}_1^1$ in such a way that each $L_n^{1,1}$ is nonmeasurable relative to the corresponding measure. For example, if $L_1^{1,1}$ corresponds to $\nu_{1,1}$, we define $V_1^1 = L_1^{1,1}$. If the measure corresponding to $L_n^{1,1}$ is $\nu_{k,i}^*$, we define $W_{k,i}'' = L_n^{1,1}$. And if the corresponding measure is $\nu_{k,i}^{00}$, we put $W_{k,i}'' = L_n^{1,1}$, but henceforth replace the measure $\nu_{k,i}^0$ by $\nu_{k,i}^- = \mu_{\mathcal{A}_k}^{L_n^{1,1}, U_i^k}$.

Suppose now that U_1^1 satisfies the conditions of Case V. Let $\mathfrak{M}_1^1$ be the family of measures whose elements are:

(1) all measures $\nu_{k,i}^*$ such that $\nu_{k,i}^*(U_1^1) = 1$;

(2) all measures $\nu_{k,i}^{00}$ such that $\nu_{k,i}^{00}(U_1^1) = 1$;

(3) $\nu_{1,1}'$.

Just as before, we consider the sets $L_n^{1,1}$. For example, suppose that $L_1^{1,1}$ corresponds to the measure $\nu_{1,1}'$. Take sets $V_1^1, W_{1,1}'' \subset L_1^{1,1}$ such that $V_1^1 \cap W_{1,1}'' = \emptyset$ and $V_1^1, W_{1,1}''$ are $\nu_{1,1}'$-nonmeasurable. It is obvious that there exist ultrafilters $\nu \ni V_1^1$, $w \ni W_{1,1}''$ such that v, t_1^1 and w, t_1^1 are two pairs of $\mathcal{A}_1$-similar ultrafilters. Therefore v, w are $\mathcal{A}_1$-similar ultrafilters. If $L_n^{1,1}$ corresponds to $\nu_{k,i}^*$ or $\nu_{k,i}^{0,0}$, we proceed as before.

Suppose now that U_1^1 satisfies the conditions of Case VI. We reason as before, except that the measure $\nu_{1,1}$ ($\nu_{1,1}'$) must be replaced by $\nu_{1,1}^0$. If $\nu_{1,1}^0$ corresponds to $L_1^{1,1}$, we define $V_1^1 = L_1^{1,1}$ and henceforth replace $\nu_{1,1}^{0,0}$ by the measure $\nu_{1,1}^{--} = \mu_{\mathcal{A}_1}^{L_1^{1,1}, U_{1,1}^0}$.

We can now proceed to the consideration of the set U_2^1 (if $m_1 > 1$). Suppose first that it satisfies the conditions of Case II. Consider the family of measures $\mathfrak{M}_2^1$ whose elements are:

(1) all measures $\nu_{k,i}^*$ such that $\nu_{k,i}^*(U_2^1) = 1$;

(2) all measures $\nu_{k,i}^{00}$ such that $\nu_{k,i}^{00}(U_2^1) = 1$;

(3) if there exists a measure $\nu_{1,1}^{--}$ and $\nu_{1,1}^{--}(U_2^1) = 1$, then $\nu_{1,1}^{--} \in \mathfrak{M}_2^1$;

(4) $\nu_{1,2}$.

It is clear that there exist pairwise disjoint sets

$$L_1^{1,2}, L_2^{1,2}, \ldots, L_n^{1,2}, \ldots \subset U_2^1,$$

which can be put in one-to-one correspondence with the measures in $\mathfrak{M}_2^1$ in such a way that each $L_n^{1,2}$ is nonmeasurable relative to the corresponding measure. For example, let the measures corresponding to $L_1^{1,2}$, $L_2^{1,2}$ be $\nu_{1,2}$, $\nu_{1,1}^{--}$, respectively. We then define $V_2^1 = L_1^{1,2}$, $W_{1,1}'' = L_2^{1,2}$. If $L_n^{1,2}$ corresponds to $\nu_{k,i}^*$, we define $W_{k,i}'' = L_n^{1,2}$. If $L_n^{1,2}$ corresponds to $\nu_{k,i}^{00}$, we define $W_{k,i}'' = L_n^{1,2}$, and henceforth replace the measure $\nu_{k,i}^0$ by $\nu_{k,i}^- = \mu_{\mathcal{A}_k}^{L_n^{1,2}, U_i^k}$.

The discussion of Case V for U_2^1 is quite natural. In Case VI one must take into consideration that either $\nu_{1,2}^0 \in \mathfrak{M}_2^1$ or $\nu_{1,2}^- \in \mathfrak{M}_2^1$.

Continuing in this way, we consider the sets $U_3^1, \ldots, U_{m_1}^1$ (if $m_1 > 2$), then the sets in the second row of the matrix, and so on. Thus, for any U_i^k satisfying the conditions of Case II, V or VI, we have constructed a set V_i^k.

If U_i^k corresponds to Case I, III, or IV, then $V_i^k = U_i^k$.

Finally, consider the set of measures $\mathfrak{M}$ whose elements are some measures $\nu_{m,p}^*, \nu_{\ell,j}^{--}$: we have $\nu_{m,p}^* \in \mathfrak{M}$ iff $U_{m,p}^* \cap \bigcup_{k,i} U_i^k = \emptyset$, and $\nu_{\ell,j}^{--} \in \mathfrak{M}$ iff $U_{\ell,j}^0 \cap \bigcup_{k,i} U_i^k = \emptyset$. If $\nu_{m,p}^* \in \mathfrak{M}$, we take a $\nu_{m,p}^*$-nonmeasurable set $W_{m,p}'' \subset U_{m,p}^*$. If $\nu_{\ell,j}^{--} \in \mathfrak{M}$, we take a $\nu_{\ell,p}^{--}$-nonmeasurable set $W_{\ell,j}'' \subset U_{\ell,j}^0$. Moreover, all the sets $W_{k,i}''$ (both those just constructed and those constructed previously) are pairwise disjoint and

$$\Theta \cap \overline{\bigcup_{k,i} W_{k,i}''} = \emptyset.$$

Thus, if U_i^k satisfies the conditions of Case IV, V or VI, we have constructed a set $W_{k,i}''$, and we can always find $\mathcal{A}_k$-similar ultrafilters v_i^k, w_i^k such that $V_i^k \in v_i^k$, $W_{k,i}'' \in w_i^k$. In

Case IV $v_i^k = s_i^k$. If U_i^k satisfies the conditions of Case I or III, then $v_i^k = s_i^k$. If U_i^k satisfies the conditions of Case II, we have an $\mathcal{A}_k$-special ultrafilter $v_i^k \ni V_i^k$. If U_i^k satisfies the conditions of Case III, then $w_i^k = t_i^k$; if $w_i^k \ni V_p^m$ (this means that $w_i^k = v_p^m$ and U_p^m satisfies the conditions of Case I, III or IV), $W_{k,i}' = V_p^m$. For every $t \in \Theta$ choose $W_t \in t$ such that all the sets V_i^k, $W_{n,j}''$, W_t form a family of pairwise disjoint sets. If U_i^k satisfies the conditions of Case III and $w_i^k = t \in \Theta$, then $W_{k,i}' = W_t$. $\square$

The following theorem is an improvement of Theorem X:

Theorem 12.2. *Consider a countable sequence of σ-algebras $\mathcal{A}_1, \ldots, \mathcal{A}_k, \ldots$ and a matrix of pairwise disjoint sets*

$$\begin{pmatrix} U_1^1 & \cdots & U_{m_1}^1 \\ \cdots\cdots\cdots\cdots \\ U_1^k & \cdots & U_{m_k}^k \\ \cdots\cdots\cdots\cdots \end{pmatrix},$$

where $m_k > \frac{3(k-1)}{2}$, $U_i^k \notin \mathcal{A}_k$. Then there exist pairwise disjoint sets $V, U_1, \ldots, U_k, \ldots$ such that if $U_k \subset Q$, $V \cap Q = \emptyset$, then $Q \notin \mathcal{A}_k$. Moreover:

(1) $U_k \subset U_{i_k}^{p_k}$;

(2) *if $k_1 \neq k_2$, then $U_{i_{k_1}}^{p_{k_1}} \neq U_{i_{k_2}}^{p_{k_2}}$;*

(3) *if $p_{k_0} \neq p_k$ for all $k \neq k_0$, then $p_{k_0} = k_0$;*

(4) *if $p_{k_1} = p_{k_2}$ $(k_1 \neq k_2)$, then $p_k \neq p_{k_1}$ for $k \notin \{k_1, k_2\}$; if $k_1 < k_2$, then $p_{k_1} = p_{k_2} = k_2$.*

Proof. We apply Theorem 12.1 and then, reasoning as in the proof of the first part of Theorem 10.5 (see also Remarks 10.5, 10.6), inductively the construct sets $V, U_1, \ldots, U_k, \ldots$. It is clear that $U_k = V_{i_k}^{p_k}$, and V is either the union of sets of the form $W_{k,i}'$ and $W_{n,j}''$ or the empty set (see Theorem 12.1). $\square$

Using Theorem 12.1 and the same reasoning as in the proof of Theorem 10.6 (see also Remarks 10.7, 10.8), we can easily prove the following refined version of Theorem XII:

Theorem 12.3. *Consider a countable sequence of σ-algebras $\mathcal{A}_1, \ldots, \mathcal{A}_k, \ldots$ and a matrix* *of pairwise disjoint sets*

$$\begin{pmatrix} U_1^1 & \cdots & U_{m_1}^1 \\ \dotfill \\ U_1^k & \cdots & U_{m_k}^k \\ \dotfill \end{pmatrix},$$

where $m_k > 3(k-1)$, $U_i^k \notin \mathcal{A}_k$. Then there exist pairwise disjoint sets $U_1, \ldots, U_k, \ldots$, $V_1, \ldots, V_k, \ldots$ such that if $U_k \subset Q$, $V_k \cap Q = \emptyset$, then $Q \notin \mathcal{A}_k$. Moreover, $U_k \subset U_{i_k}^k$.

Using our methods, it is rather easy to prove the following theorem, which should be compared with Theorems 10.7 and 12.2:

Theorem 12.4. *Consider a countable sequence of σ-algebras $\mathcal{A}_1, \ldots, \mathcal{A}_k, \ldots$ and a matrix of pairwise disjoint sets*

$$\begin{pmatrix} U_1^1 & \cdots & U_{m_1}^1 \\ \dotfill \\ U_1^k & \cdots & U_{m_k}^k \\ \dotfill \end{pmatrix},$$

where $m_k > 2(k-1)$, $U_i^k \notin \mathcal{A}_k$. Then there exist pairwise disjoint sets $V, U_1, \ldots, U_k, \ldots$ such that if $U_k \subset Q$, $V \cap Q = \emptyset$, then $Q \notin \mathcal{A}_k$. Moreover, $U_k \subset U_{i_k}^k$.

Up to now we have refined all the main results of this memoir, with the exception of Theorems VI and VIII. We now address ourselves to the task of refining these two theorems.

Let $\mathcal{A}_1, \ldots, \mathcal{A}_k, \ldots$ be a countable sequence of σ-algebras such that for every k there exist pairwise disjoint sets $Q_1^k, \ldots, Q_{m_k}^k \notin \mathcal{A}_k$. For every Q_i^k, there are four possibilities:

(1) there exists an irregular point $s_i^k \in \overline{Q_i^k}$, and s_i^k is an $\mathcal{A}_k$-special ultrafilter;

(2) possibility (1) does not hold, but there exist irregular points s_i^k, t_i^k such that $s_i^k \in \overline{Q_i^k}$, $t_i^k \notin \overline{Q_i^k}$, and s_i^k, t_i^k are $\mathcal{A}_k$-similar ultrafilters;

(3) neither of the previous possibilities hold, but there exists a pair of $\mathcal{A}_k$-similar ultrafilters such that one of them (denote it by s_i^k) contains Q_i^k and is an irregular point, while the other does not contain Q_i^k;

(4) none of (1), (2), (3) holds.

Denote the collection of all irregular points of the type s_i^k and t_i^k, which we have considered up to now, by Z. If (3) holds for Q_i^k, we consider a set $Q_{k,i}^2$ and an ultrafilter t_i^k such that $Q_{k,i}^2 \in t_i^k$, and s_i^k, t_i^k are $\mathcal{A}_k$-similar ultrafilters. If (4) holds for Q_i^k, we consider a set $Q^1_{k,i} \subset Q^k_i$. There are two possibilities for $Q^1_{k,i}$:

(a) $\overline{Q^1_{k,i}} \cap \ker \mathcal{A}_k$ contains an $\mathcal{A}_k$-special ultrafilter s^k_i;

(b) condition (a) is not satisfied, but there exist a set $Q^2_{k,i}$ and $\mathcal{A}_k$-similar ultrafilters s^k_i, t^k_i such that $Q^1_{k,i} \in s^k_i$, $Q^2_{k,i} \in t^k_i$.

On the strength of the arguments presented in the previous and present sections, we may demand that

$$Q^j_{k,i} \cap Q^{j'}_{k',i'} \neq \emptyset \text{ iff } k = k', \ i = i', \ j = j'; \quad Z \cap \overline{\bigcup_{k,i,j} Q^j_{k,i}} = \emptyset.$$

In view of this construction, in conjunction with Theorems 10.2, 10.4 and Remarks 10.2, 10.4, we arrive at the following refined versions of Theorems VI and VIII:

Theorem 12.5. *Let $\mathcal{A}_1, \ldots, \mathcal{A}_k, \ldots$ be a countable sequence of σ-algebras. Suppose that for each k we are given a finite sequence of pairwise disjoint sets $Q^k_1, \ldots, Q^k_{m_k} \notin \mathcal{A}_k$, where $m_k > \frac{5}{2}(k-1)$ if $k \neq 2$. Suppose, moreover, that there exist three pairwise disjoint sets $Q^{\alpha_1}_{\beta_1}, Q^{\alpha_2}_{\beta_2}, Q^{\alpha_3}_{\beta_3}$, where each of the superscripts α_i is either 1 or 2 and two of them are different. Then for each set Q^k_i one can construct a set $\hat{Q}^k_i \subset Q^k_i$ such that $\hat{Q}^k_i \notin \mathcal{A}_k$; either $\hat{Q}^k_i = \hat{Q}^\ell_j$ or $\hat{Q}^k_i \cap \hat{Q}^\ell_j = \emptyset$; and there exist pairwise disjoint sets $V, \hat{Q}^{\gamma_1}_{i_1}, \ldots, \hat{Q}^{\gamma_k}_{i_k}, \ldots$ such that if $\hat{Q}^{\gamma_k}_{i_k} \subset Q$, $V \cap Q = \emptyset$, then $Q \notin \mathcal{A}_k$.*

Theorem 12.6. *Consider a countable sequence of σ-algebras $\mathcal{A}_1, \ldots, \mathcal{A}_k, \ldots$ such that for every k there exists a finite sequence of pairwise disjoint sets $Q^k_1, \ldots, Q^k_{m_k} \notin \mathcal{A}_k$, where $m_k > 4(k-1)$. Then for each set Q^k_i one can construct a set $\hat{Q}^k_i \subset Q^k_i$ such that $\hat{Q}^k_i \notin \mathcal{A}_k$; either $\hat{Q}^k_i = \hat{Q}^\ell_j$ or $\hat{Q}^k_i \cap \hat{Q}^\ell_j = \emptyset$; and there exist pairwise disjoint sets $\hat{Q}^1_{i_1}, \ldots, \hat{Q}^k_{i_k}, \ldots, V_1, \ldots, V_k, \ldots$ such that if $\hat{Q}^k_{i_k} \subset Q$, $V_k \cap Q = \emptyset$, then $Q \notin \mathcal{A}_k$.*

13. Sets not belonging to Semi-Lattices of Subsets and not belonging to Lattices of Subsets

In this memoir we have repeatedly encountered the following situation: we have considered a sequence of algebras $\mathcal{A}_1, \ldots, \mathcal{A}_k, \ldots$, and constructed a sequence of pairwise disjoint sets $U_1, \ldots, U_k, \ldots$ such that $U_k \notin \mathcal{A}_k$. Hence the following question is quite natural: under what conditions, imposed on a finite (countable) sequence of algebras (almost σ-algebras), does there exist a sequence of corresponding sets $U_1, \ldots, U_k, \ldots$? The technique we have developed here enables us to answer this question quite easily. But it may perhaps be of some interest that the solution can be found by considering not algebras but rather only semi-lattices of subsets.

Definition 13.1. A set $\mathcal{A} \subset \mathcal{P}(X)$ is called a semi-lattice if $M_1, M_2 \in \mathcal{A}$ implies that $M_1 \cup M_2 \in \mathcal{A}$.

If $\emptyset \notin \mathcal{A}$, then $\mathcal{A} \cup \{\emptyset\}$ is, obviously, a semi-lattice. Therefore, we shall assume, for simplicity, that if $\mathcal{A}$ is a semi-lattice, then $\emptyset \in \mathcal{A}$.

Let $Q \notin \mathcal{A}$. Then there exists an ultrafilter $s \ni Q$ such that

$$(s \cap Q) \cap \mathcal{A} = \emptyset.$$

¿From this simple assertion one easily deduces the following

Theorem 13.1. *If $\mathcal{A}_1, \ldots, \mathcal{A}_n$ is a finite sequence of semi-lattices and for every k, $1 \le k \le n$, there exist more than $k-1$ pairwise disjoint sets which are not member of $\mathcal{A}_k$, then there exists a sequence of pairwise disjoint sets $Q_1, \ldots, Q_n$ such that $Q_k \notin \mathcal{A}_k$, $1 \le k \le n$.*

It is very easy to prove that the bound $k - 1$ is best possible.

To every semi-lattice $\mathcal{A}$ one can associate a set of ultrafilters called the kernel of $\mathcal{A}$ and denoted by $\mathrm{kr}\mathcal{A}$. By definition, $s \in \mathrm{kr}\mathcal{A}$ iff there exists a set $Q \in s$ such that

$$(s \cap Q) \cap \mathcal{A} = \emptyset.$$

Lemma 13.1. *If $\mathcal{A}$ is an algebra, then $\ker \mathcal{A} = \mathrm{kr}\mathcal{A}$.*

Proof. Obviously, $\ker \mathcal{A} \subset \mathrm{kr}\mathcal{A}$. On the other hand, let $s \in \mathrm{kr}\mathcal{A}$. If $M \notin \mathcal{A}$ for any $M \in s$, then s is an $\mathcal{A}$-special ultrafilter, and $s \in \ker \mathcal{A}$. Let

$$Q \in s,$$
$$(s \cap Q) \cap \mathcal{A} = \emptyset,$$
$$Q \subset D \in \mathcal{A}.$$

Then there exists $t \in \overline{D \backslash Q}$ such that s, t are $\mathcal{A}$-similar, and $s \in \ker \mathcal{A}$. $\square$

The concept of an almost σ-semi-lattice is defined quite naturally.

Let $\mathcal{A}$ be an almost σ-semi-lattice and $\mathcal{A} \neq \mathcal{P}(X)$. We define a σ-additive measure μ on X which assumes the values 0 and 1: $\mu(M) = 0$ iff $\mathcal{P}(M) \subset \mathcal{A}$; $\mu(M) = 1$ iff $\mu(X \backslash M) = 0$. It is obvious that the kernel of the σ-algebra of all μ-measurable sets is precisely $\overline{\ker \mathcal{A}}$.[43]

Now, using the reasoning from Section 5, one readily proves the following

Theorem 13.2. *If $\mathcal{A}_1, \ldots, \mathcal{A}_k, \ldots$ is a countable sequence of almost σ-semi-lattices and for every k there exist more than $k - 1$ pairwise disjoint sets which are not members of $\mathcal{A}_k$, then there exists a sequence of pairwise disjoint sets $Q_1, \ldots, Q_k, \ldots$ such that $Q_k \notin \mathcal{A}_k$.*

After considering semi-lattices of subsets, we proceed to the following natural

Definition 13.2. A set $\mathcal{A} \subset \mathcal{P}(X)$ is called a lattice if $M_1, M_2 \in \mathcal{A}$ implies $\mathcal{A} \ni M_1 \cup M_2$, $M_1 \cap M_2$.

We shall assume, for simplicity, that if $\mathcal{A}$ is a lattice, then $\emptyset \in \mathcal{A}$. When we speak of a kernel $\ker \mathcal{A}$ of a lattice $\mathcal{A}$, we consider $\mathcal{A}$ as a semi-lattice.

Let us now proceed to two statements about lattices. ¿From the proof of Proposition 5.1, we can obtain the following proposition (which is a generalization of Proposition 5.1):

Proposition 13.1. *Consider a countable sequence of lattices $\mathcal{A}_1, \ldots, \mathcal{A}_k, \ldots$. Assume that there is a matrix*

$$\begin{pmatrix} U_1^1 & & & \\ U_1^2 & U_1^2 & U_3^2 & \\ \cdots\cdots\cdots\cdots\cdots\cdots\cdots \\ U_1^k \cdots\cdots\cdots U_k^k & U_{k+1}^k \\ \cdots\cdots\cdots\cdots\cdots\cdots\cdots \end{pmatrix}$$

of pairwise disjoint sets such that $U_i^k \notin \mathcal{A}_k$. (The first row contains one set, and the k-th row ($k > 1$) contain $k + 1$ sets.) Then there exists a set $U = \bigcup_k U_{i_k}^k$ such that

---

[43] Let $|X| \geq \aleph_0$. Define a semi-lattice $\mathcal{A}$ as follows: $M \notin \mathcal{A}$ iff $|M| = 1$. Obviously,

$$\ker \mathcal{A} = X \neq \overline{X} = \beta X.$$

In this memoir we have not gone into the question of whether, or under what conditions $\ker \mathcal{A}$ (where $\mathcal{A}$ is, of course, an algebra) is a closed set. To justify this omission we will only point out that this question is not related to our main topic. We mention that in Section 5 (footnote 20) it is proved that $\ker \mathcal{A}$ is a countably compact subspace of βX.

(a) $\mathcal{A}_k \not\ni \bigcup_{m \geq k} U_{im}^m$;

(b) if $U \in \mathcal{A}_k$, then $k > 1$ and $\mathcal{A}_k \ni \bigcup_{m < k} U_{im}^m$.

Definition 13.3. A lattice $\mathcal{A}$ is called a σ-lattice if, for any countable sequence $\{M_k\} \subset \mathcal{A}$, it is true that $\mathcal{A} \ni \bigcup_k M_k, \bigcap_k M_k$.

The following theorem should be compared with Corollary 6.1 and with the improvement of this corollary in Remark 6.1:

Theorem 13.3. *Let* $|X| = 2^{\aleph_0}$, *and let* $\mathcal{A}_1, \ldots \mathcal{A}_k, \ldots$ *be a countable sequence of σ-lattices none of which is all of* $\mathcal{P}(X)$, *and assume that* $\{x\} \in \mathcal{A}_k$, *if* $x \in X$. *Then there is a matrix of sets*

$$
\begin{pmatrix}
V_1^1 \supset & V_2^1 \supset & V_3^1 \supset & V_4^1 & \supset \ldots \supset & V_k^1 & \supset & V_{k+1}^1 & \supset & V_{k+2}^1 & \supset \ldots \supset & V_j^1 \supset \ldots \\
V_1^2 \supset & & V_3^2 \supset & V_4^2 & \supset \ldots \supset & V_k^2 & \supset & V_{k+1}^2 & \supset & V_{k+2}^2 & \supset \ldots \supset & V_j^2 \supset \ldots \\
V_1^3 \supset & & & V_4^3 & \supset \ldots \supset & V_k^3 & \supset & V_{k+1}^3 & \supset & V_{k+2}^3 & \supset \ldots \supset & V_j^3 \supset \ldots \\
& & & & \cdots\cdots\cdots\cdots & & & & & & & \\
V_1^k \supset & & & & & & & V_{k+1}^k & \supset & V_{k+2}^k & \supset \ldots \supset & V_j^k \supset \ldots \\
& & & & \cdots\cdots\cdots\cdots & & & & & & &
\end{pmatrix}
$$

such that $V_j^k \notin \mathcal{A}_k$, $V_1^k \cap V_1^m = \emptyset$ $(k \neq m)$, $V_i^k \supset V_j^k$ $(i < j)$, *and*

(a) $(\bigcup_{i \geq k} V_1^i) \cup (\bigcup_{i < k} V_k^i) \notin \mathcal{A}_k$ *(if* $k = 1$, *then* $\bigcup_{i < k} V_k^i = \emptyset$);

(b) $\mathcal{A}_k \not\ni \bigcup_{i \geq k} V_1^i$;

(c) *if* $\mathcal{A}_k \ni \bigcup_i V_1^i$, *then* $k > 1$ *and* $\mathcal{A}_k \ni \bigcup_{i < k} V_1^i$.

Proof. Obviously, $\mathrm{kr}\mathcal{A}_k \subset \beta X \backslash X$, and it is rather easy to prove that

$$|\overline{Q} \cap \mathrm{kr}\mathcal{A}_k| \geq \aleph_0$$

if $Q \notin \mathcal{A}_k$. As in the proof of Corollary 6.1 (see also Theorem 13.2) we can take pairwise disjoint sets $Q_1, \ldots, Q_k, \ldots$ such that $Q_k \notin \mathcal{A}_k$, and construct a matrix of pairwise disjoint sets

$$
\begin{pmatrix}
U_1^1 & & & \\
U_1^2 & U_2^2 & U_3^2 & \\
\cdots\cdots\cdots\cdots\cdots\cdots\cdots\cdots \\
U_1^k \cdots\cdots\cdots & & U_k^k & U_{k+1}^k \\
\cdots\cdots\cdots\cdots\cdots\cdots\cdots\cdots
\end{pmatrix}
$$

such that $U_i^k \notin \mathcal{A}_k$. By Proposition 13.1, we construct a corresponding sequence of sets

$$U_{i_1}^1, U_{i_2}^2, \ldots, U_{i_k}^k, \ldots .$$

Let $V_1^k = U_{i_k}^k$. For any k there exists a countable sequence of sets

$$W_1^k \supset W_2^k \supset \ldots \supset W_\ell^k \supset \ldots$$

such that $V_1^k \supset W_1^k$, $W_\ell^k \notin \mathcal{A}_k$, $\bigcap_\ell W_\ell^k = \emptyset$. Therefore, there exists a sequence of natural numbers

$$\ell_2 > \ell_3 > \ldots > \ell_k > \ldots$$

such that for $k > 1$

$$\left(\bigcup_{i \geq k} V_1^i\right) \cup \left(\bigcup_{i < k} W_{\ell_k}^i\right) \notin \mathcal{A}_k.$$

It remains only to denote

$$V_j^k = W_{\ell_j}^k$$

for $j > 1$ and $k < j$. $\square$

Remark 13.1. In the proof of Theorem 13.3 we did not fully use the property of σ-lattices that if $\aleph_0$ sets $M_1, \ldots, M_i, \ldots \in \mathcal{A}_k$, then $\mathcal{A}_k \ni \bigcup_i M_i$. We use only the fact that these lattices are almost σ-semi-lattices, and if $\aleph_0$ sets $M_1, \ldots, M_i, \ldots \in \mathcal{A}_k$, then $\mathcal{A}_k \ni \bigcap_i M_i$.

14. UNSOLVED PROBLEMS

Problem 1. An interesting regularity can be detected in the results formulated in Section 2. Assertions valid for a finite sequence of algebras (Corollary 2.1, the first parts of Theorems V, VII, and Theorems IX, XI) are also valid for a countable sequence of σ-algebras (the first part of Theorem III, and Theorems VI, VIII, X, XII, respectively). However, the statement of the first part of Theorem I for a finite sequence of algebras turns out to be true not only for a countable sequence of σ-algebras but also for a countable sequence of almost σ-algebras (Theorem II). It is therefore natural to ask: Can the σ-additivity condition imposed on the algebras in the first part of Theorem III, and Theorems VI, VIII, X, XII (or in parts of these theorems) be weakened in some way?

The following two problems were mentioned at the end of Section 2.

Problem 2. Theorem IX is apparently true if $m_k > k - 1$. Then Theorem X is also true with the same bound.

We shall prove that if our hypothesis is true, this bound is best possible.

Proposition 14.1. *For any natural number $n > 1$, one can construct algebras $\mathcal{A}_1, \ldots, \mathcal{A}_n$ for which the statement of Theorem IX is false, and there exists a matrix of pairwise disjoint sets*

$$
\begin{pmatrix}
U_1^1 & & \\
\cdots\cdots\cdots\cdots & & \\
U_1^k & \cdots & U_{m_k}^k \\
\cdots\cdots\cdots\cdots & & \\
U_1^n & \cdots & U_{n-1}^n
\end{pmatrix},
$$

$m_k = k$ for $k < n$, $m_n = n - 1$, $U_i^k \notin \mathcal{A}_k$.

Proof. We demonstrate the construction for $n = 5$. For all k, $X \in \mathcal{A}_k$, and

$$
\{a_i^k\} = \ker \mathcal{A}_k \cap \overline{U_i^k}
$$

for any of the sets U_i^k. The $\mathcal{A}_k$-similar sets, $k = 1, \ldots, 5$, are as follows:

 for $k = 1:$ $\{a_1^1, a_1^2\}$;

 for $k = 2:$ $\{a_1^2, a_2^2\}$;

 for $k = 3:$ $\{a_1^3, a_1^1\}, \{a_2^3, a_2^2\}, \{a_3^3, a_3^4\}$;

 for $k = 4:$ $\{a_1^4, a_1^1\}, \{a_2^4, a_2^2\}, \{a_3^4, a_4^4\}$;

 for $k = 5:$ $\{a_1^5, a_1^1\}, \{a_2^5, a_2^2\}, \{a_3^5, a_3^3\}, \{a_4^5, a_4^4\}$.

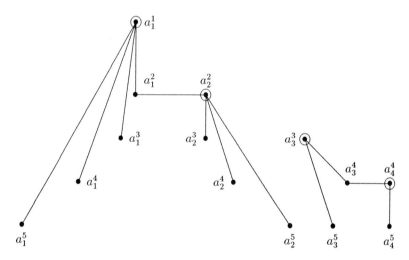

The statement of Theorem IX is true for $\mathcal{A}_1, \mathcal{A}_2, \mathcal{A}_3, \mathcal{A}_4$. Moreover, it is necessary that

$$
U_1 \in a_1^1, \ U_2 \in a_2^2, \ U_3 \in a_3^3, \ U_4 \in a_4^4
$$

(the sets U_k are taken from the formulation of Theorem IX). Hence it follows that the statements of Theorem IX is false for $\mathcal{A}_1, \mathcal{A}_2, \mathcal{A}_3, \mathcal{A}_4, \mathcal{A}_5$.

By analogy, such algebras can be constructed for any $n > 1$. $\square$

Problem 3. Theorem XI is apparently true if $m_1 > 0$, and $m_k > 2k - 1$ $(k > 1)$. Then Theorem XII is also true with the same bound.

We shall prove that if our hypothesis is true, this bound is best possible.

Proposition 14.2. *For any natural number $n > 1$, one can construct algebras $\mathcal{A}_1, \ldots, \mathcal{A}_n$ for which the statement of Theorem XI is false, and there exists a matrix of pairwise disjoint sets*

$$
\begin{pmatrix}
U_1^1 & & \\
\cdots\cdots\cdots\cdots\cdots & & \\
U_1^k & \cdots & U_{m_k}^k \\
\cdots\cdots\cdots\cdots\cdots & & \\
U_1^n & \cdots & U_{2n-1}^n
\end{pmatrix},
$$

$m_1 = 1$, $m_k = 2k$ for $1 < k < n$, $m_n = 2n - 1$, $U_i^k \notin \mathcal{A}_k$.

Proof. We demonstrate the construction for $n = 5$. For all k, $X \in \mathcal{A}_k$, and

$$
\{a_i^k\} = \ker \mathcal{A}_k \cap \overline{U_i^k}
$$

for any of the sets U_i^k. The $\mathcal{A}_k$-similar sets, $k = 1, \ldots, 5$, are as follows:

for $k = 1$: $\{a_1^1, a_1^2\}$;

for $k = 2$: $\{a_1^2, a_2^2\}, \{a_3^2, a_1^1\}, \{a_4^2, a_3^3\}$;

for $k = 3$: $\{a_1^3, a_1^1\}, \{a_2^3, a_1^2\}, \{a_3^3, a_4^3\}, \{a_5^3, a_4^2\}, \{a_6^3, a_5^4\}$;

for $k = 4$: $\{a_1^4, a_1^1\}, \{a_2^4, a_1^2\}, \{a_3^4, a_4^2\}, \{a_4^4, a_3^3\}, \{a_5^4, a_6^4\}, \{a_7^4, a_6^3\}, \{a_8^4, a_7^5\}$;

for $k = 5$: $\{a_1^5, a_1^1\}, \{a_2^5, a_1^2\}, \{a_3^5, a_4^2\}, \{a_4^5, a_3^2\}, \{a_5^5, a_6^3\}, \{a_6^5, a_5^4\}, \{a_7^5, a_8^4\}, \{a_9^5, a_8^4\}.$

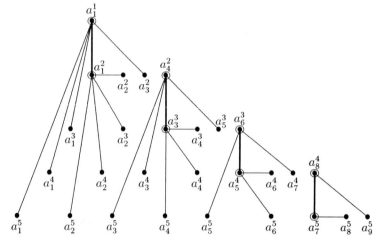

The statement of Theorem XI is true for $\mathcal{A}_1, \mathcal{A}_2, \mathcal{A}_3, \mathcal{A}_4$. Moreover, it is necessary that

$$a_1^1, a_1^2 \in \overline{U}_1 \cup \overline{V}_1;$$

$$a_4^2, a_3^3 \in \overline{U}_2 \cup \overline{V}_2;$$

$$a_6^3, a_5^4 \in \overline{U}_3 \cup \overline{V}_3;$$

$$a_8^4, a_7^5 \in \overline{U}_4 \cup \overline{V}_4$$

(the sets U_k, V_k are taken from the formulation of Theorem XI). Hence it follows that the statement of Theorem XI is false for $\mathcal{A}_1, \mathcal{A}_2, \mathcal{A}_3, \mathcal{A}_4, \mathcal{A}_5$.

By analogy, such algebras can be constructed for any $n > 1$. $\square$

We note that Problem 1 has a set-theoretic character, whereas Problems 2 and 3 are combinatorial in nature.

ADDED IN PROOF

Remark A. One idea in the proof of Lemma 7.5 yields an elegant proof of Theorem 4.2, using the first part of Theorem I. We use the same notation as in the proof of Theorem 4.2. For every k, $1 \leq k \leq n$, consider the nonempty set

$$Z_k^* \subset \bigcup_{k,i} \{s_i^k\}$$

which has the following properties:

(a) $|Z_k^*| > \frac{4}{3}(k-1)$ if $k \neq 2$;

(b) every $z \in Z_k^*$ satisfies one of the following three conditions:

(b_1) z is $\mathcal{A}_k$-special ultrafilter;

(b_2) z has an $\mathcal{A}_k$-similar ultrafilter in Z_k^*;

(b_3) if z satisfies neither (b_1) nor (b_2), then it has an $\mathcal{A}_k$-similar ultrafilter in $\beta X \smallsetminus \bigcup_{k,i}\{s_i^k\}$.

Consider the algebra $\mathcal{A}_k^0$. We define an ultrafilder z to be $\mathcal{A}_k^0$-special if $z \in Z_k^*$ and it satisfies either (b_1) or (b_3). Similarly, ultrafilters a, b ($a \neq b$) will be called $\mathcal{A}_k^0$-similar if $a, b \in Z_k^*$ and they are $\mathcal{A}_k$-similar. By the first part of Theorem I, given the algebras $\mathcal{A}_1^0, \ldots, \mathcal{A}_n^0$, we can construct sets of ultrafilters S_n^*, T_n^* (to avoid confusion, we are not denoting these sets by S_n, T_n). Clearly, we can put $S_n = S_n^*$.

Similarly, we can give an elegant proof of Theorem 10.2, using the first part of Theorem V.

Remark B. Problems 2 and 3 have now been solved. The solutions will be published in the second part of this memoir, now in preparation. At the same time, we will also substantially strengthen Theorems 10.7, 12.4.

REFERENCES

[BD] J. Baumgartner and E. van Douwen, *Strong real compactness and weakly measurable cardinals*, Topology and its Applications **35** (1990), 239–251.

[E] R. Engelking, *General Topology*, 1977, PWN, (Manuscript of the second edition, 1985).

[Er] P. Erdős, *Some remarks on set theory*, Proc. A.M.S. (1950), 127–141.

[G] E. Grzegorek, *On saturated sets of Boolean rings and Ulam's problem on sets of measures*, Fund. Math. **CX** (1980), 153–161.

[GS] M. Gitik and S. Shelah, *Forcing with ideals and simple forcing notions*, Israel J. Math. **68** (1989), 129–160.

[K] A. Kamburelis, *A new proof of the Gitik-Shelah theorem*, Israel J. Math. **72** (1990), 373–380.

[M] S. Mrówka, *On the potency of βN*, Coll. Math. **7** (1959-60), 23–25.

[P] B. Pospíšil, *Remark on bicompact spaces*, Ann. of Math. **38** (1937), 845–846.

[S] S. Shelah, *Iterated forcing and normal ideals on ω_1*, Israel J. Math. **60** (1987), 345–380.

[Si] R. Sikorski, *Boolean Algebras*, Springer-Verlag, 1964.

[So] R. Solovay, *Real-valued measurable cardinals*, Axiomatic Set Theory, Proc. Symp. Pure Math. **13(I)** (1971), American Mathematical Society, Rhode Island, 397–428.

[U] S. Ulam, *Zur Masstheorie in der allgemeinen Mengenlehre*, Fund. Math. **XVI** (1930), 140-150.

21 Neufeld Street, Apt. #13, Kiryat Herzog, Bnei Brak 51242, Israel

Editorial Information

To be published in the *Memoirs*, a paper must be correct, new, nontrivial, and significant. Further, it must be well written and of interest to a substantial number of mathematicians. Piecemeal results, such as an inconclusive step toward an unproved major theorem or a minor variation on a known result, are in general not acceptable for publication. *Transactions* Editors shall solicit and encourage publication of worthy papers. Papers appearing in *Memoirs* are generally longer than those appearing in *Transactions* with which it shares an editorial committee.

As of September 1, 1992, the backlog for this journal was approximately 9 volumes. This estimate is the result of dividing the number of manuscripts for this journal in the Providence office that have not yet gone to the printer on the above date by the average number of monographs per volume over the previous twelve months. (There are 6 volumes per year, each containing about 3 or 4 numbers.)

A Copyright Transfer Agreement is required before a paper will be published in this journal. By submitting a paper to this journal, authors certify that the manuscript has not been submitted to nor is it under consideration for publication by another journal, conference proceedings, or similar publication.

Information for Authors

Memoirs are printed by photo-offset from camera copy fully prepared by the author. This means that the finished book will look exactly like the copy submitted.

The paper must contain a *descriptive title* and an *abstract* that summarizes the article in language suitable for workers in the general field (algebra, analysis, etc.). The *descriptive title* should be short, but informative; useless or vague phrases such as "some remarks about" or "concerning" should be avoided. The *abstract* should be at least one complete sentence, and at most 300 words. Included with the footnotes to the paper, there should be the 1991 *Mathematics Subject Classification* representing the primary and secondary subjects of the article. This may be followed by a list of *key words and phrases* describing the subject matter of the article and taken from it. A list of the numbers may be found in the annual index of *Mathematical Reviews*, published with the December issue starting in 1990, as well as from the electronic service e-MATH [**telnet e-MATH.ams.org** (or **telnet 130.44.1.100**). Login and password are **e-math**]. For journal abbreviations used in bibliographies, see the list of serials in the latest *Mathematical Reviews* annual index. When the manuscript is submitted, authors should supply the editor with electronic addresses if available. These will be printed after the postal address at the end of each article.

Electronically-prepared manuscripts. The AMS encourages submission of electronically-prepared manuscripts in $\mathcal{A}_{\mathcal{M}}\mathcal{S}$-TEX or $\mathcal{A}_{\mathcal{M}}\mathcal{S}$-LATEX. To this end, the Society has prepared "preprint" style files, specifically the amsppt style of $\mathcal{A}_{\mathcal{M}}\mathcal{S}$-TEX and the amsart style of $\mathcal{A}_{\mathcal{M}}\mathcal{S}$-LATEX, which will simplify the work of authors and of the production staff. Those authors who make use of these style files from the beginning of the writing process will further reduce their own effort.

Guidelines for Preparing Electronic Manuscripts provide additional assistance and are available for use with either $\mathcal{A}_{\mathcal{M}}S$-TEX or $\mathcal{A}_{\mathcal{M}}S$-LATEX. Authors with FTP access may obtain these *Guidelines* from the Society's Internet node e-MATH.ams.org (130.44.1.100). For those without FTP access they can be obtained free of charge from the e-mail address guide-elec@math.ams.org (Internet) or from the Publications Department, P. O. Box 6248, Providence, RI 02940-6248. When requesting *Guidelines* please specify which version you want.

Electronic manuscripts should be sent to the Providence office only after the paper has been accepted for publication. Please send electronically prepared manuscript files via e-mail to pub-submit@math.ams.org (Internet) or on diskettes to the Publications Department address listed above. When submitting electronic manuscripts please be sure to include a message indicating in which publication the paper has been accepted.

For papers not prepared electronically, model paper may be obtained free of charge from the Editorial Department at the address below.

Two copies of the paper should be sent directly to the appropriate Editor and the author should keep one copy. At that time authors should indicate if the paper has been prepared using $\mathcal{A}_{\mathcal{M}}S$-TEX or $\mathcal{A}_{\mathcal{M}}S$-LATEX. The *Guide for Authors of Memoirs* gives detailed information on preparing papers for *Memoirs* and may be obtained free of charge from AMS, Editorial Department, P.O. Box 6248, Providence, RI 02940-6248. The *Manual for Authors of Mathematical Papers* should be consulted for symbols and style conventions. The *Manual* may be obtained free of charge from the e-mail address cust-serv@math.ams.org or from the Customer Services Department, at the address above.

Any inquiries concerning a paper that has been accepted for publication should be sent directly to the Editorial Department, American Mathematical Society, P. O. Box 6248, Providence, RI 02940-6248.